PHYSICS AND TECHNOLOGY OF SILICON CARBIDE DEVICES

PHYSICS AND TECHNOLOGY OF SILICON CARBIDE DEVICES

Editor

George Gibbs

scitus academics

Physics and Technology of Silicon Carbide Devices
Edited by **George Gibbs**

ISBN: 978-1-68117-643-7
Library of Congress Control Number: 2016946504

SCITUS Academics LLC,
616, Corporate Way, Suite 2, 4766,
Valley Cottage, NY 10989

www.scitusacademics.com

Notice

Printed in United States of America on Acid Free Paper

PREFACE

Silicon Carbide Devices is definitely the most popular semiconductor material for power devices. On the contrary, Si-based power equipment approaches its material limits, which have led to many attempts to find alternatives to Si-based power equipment for better performance. With the rapid developments and innovations in the semiconductor industry, Silicon Carbide (SiC) power plants have moved from immature prototypes in laboratories to a practical substitute for Si-based power plants in high-performance and high-efficiency applications. SiC devices have many convincing advantages-high breakdown voltage, high frequency switching, high operating temperature, high electric field and low losses. Silicon Carbide (SiC) equipment belongs to the so-called broad-band semiconductor group, which offers a selection of attractive features for high-voltage power semiconductors compared to commonly used silicon (Si). Some SiC power devices such as Schottky barrier diodes (SBDs), metal-oxide-semiconductor field effect transistors (MOSFETs), FET junctions (JFETs) and their integrated modules have recently been launched on the market.

Physics and Technology of Silicon Carbide Devices covers up-to-date methods and techniques in the manufacturing, modeling, characterization, processing technologies, etc. for Carbide Silicon Devices.

TABLE OF CONTENTS

1

ION SYNTHESIS OF SIC AND ITS INSTABILITY AT HIGH TEMPERATURES

Kair Kh. Nussupov[1] and Nurzhan B. Beisenkhanov[1]

[1] Kazakh-British Technical University, Kazakhstan

1. INTRODUCTION

As is known, such advantages of silicon carbide as a high hardness (4th place after diamond) [36, 37], high chemical and radiation resistance, high melting point, etc. became the basis of its wide application not only in microelectronics [1], but also as refractory and abrasive materials. Silicon carbide is included in the oxidation resistant composite materials [42] used in coating system for "Space Shuttle", capable of withstanding temperatures up to 1500°C at the entrance of the ship into the atmosphere. In many attempts to develop an effective oxidation-protection coating for carbon-carbon composites with excellent mechanical properties at elevated temperature, silicon carbide coating has shown the best performance for short periods of up to 1900K [39]. For longer periods and higher temperature applications, a challenging coating system should be developed.

Silicon carbide is regarded by researchers as a suitable material for the front wall structures of fusion reactors. The boers, cutting disks, grinding paper of SiC can be used for boring, drilling, surface grinding and cutting of steel, nonferrous metals, natural stone, concrete, wood and plastic.

The stability of silicon carbide to high temperature treatment is of special interest. As a special application, silicon carbide can be thermally oxidized in the form of SiO_2, and the devices which can be easily fabricated on Si substrate (Power MOSFET, IGBT, MOS controlled thyristor, etc.) can also be fabricated on SiC substrate [23]. In paper [23] the parabolic growth of thickness of thermal oxide versus oxidation time was observed, and the slope of the plots increases with increasing temperature. The thickness values of oxide films were about 23-77 nm (Si-face, wet oxidation), 18-63 nm (Si-face, dry oxidation), 210-810 nm (C-face, wet oxidation) and 125-260 nm (C-face, dry oxidation) for oxidation time 6 h and depend on temperature value (1000, 1050, 1110 or 1150°C).

Doped with different impurities, silicon carbide is used in semiconductor technology [63, 12].

Field-effect transistors, diodes and other electronic devices based on SiC have several advantages compared to similar silicon devices, for example, the opportunity to work at temperatures up to 600°C, high speed and high radiation resistance. A large number of polytypes of SiC makes it possible to create heteropolytype structures [31, 32, 33]. Currently, using the methods of vacuum sublimation [48], molecular beam epitaxy [15], the epitaxial and heteropolytype layers based on the cubic 3C-SiC and two hexagonal 6H-SiC, 4H-SiC on substrates of SiC, are grown. Heteroepitaxial layers of 3C-SiC on substrates of Si by chemical vapor deposition (CVD) [41] are grown. At the temperatures below 1200°C there are conditions for the growth of both poly- and nanocrystalline SiC with different degrees of crystallinity and structure of the cubic polytype 3C-SiC. Such conditions were realized in the magnetron sputtering [25, 56], laser ablation [53] and plasma deposition [36], plasma-enhanced chemical vapor deposition [19, 43], molecular beam epitaxy [16]. At temperatures below 1500°C in the direct deposition of carbon and silicon ions with energy of ~100 eV, the growth of nanocrystalline films with a consistent set of the polytypes 3C, 21R, 27R, 51R, 6H is possible [49, 50, 51].

In recent years there has been an intensification of studies on the synthesis of SiC by high-dose carbon ion implantation into Si [37, 35]. In addition, the synthesis of SiC by high dose implantation of carbon ions into silicon is also of fundamental scientific interest due to the wide practical application [9 - 11,47], for example, to create a coating and insulating SiC layers in the manufacture of integrated circuits. High quality crystalline β-SiC film on SiO_2 can be obtained by multiple implantations of carbon ions into silicon and subsequent selective oxidation of the top layer of Si [52]. Intensively developing area is the formation by this technique in SiO_2 of nanostructured systems with inclusions of nanocrystals and clusters of Si, SiC and C, providing at the expense of size effects luminescence throughout the visible spectrum [57]. The study of the stability of these films to high temperature treatment is also of special interest.

The ion synthesis of silicon carbide and studies of crystallization process attract attention of researchers [12, 35, 37, 63]. The implantation of single energy carbon ions with a Gaussian distribution on the depth in silicon is of interest due to a presence of wide range of nanolayers with different concentrations of carbon and silicon atoms and, therefore, a presence of different clusters and nanocrystals of silicon, carbon and silicon carbide in the implanted layer after implantation and annealing. The properties of these layers have been investigated in detail. In previous investigations, the carbon ions with energy of 40 keV were used for considered purposes in a number of papers [8, 13,20], and doses ranged 10^{16}-10^{18} cm^{-2} were used in almost all of investigations, when the ion synthesis of a silicon carbide film was carried out [2]. The IR absorption technique was widely applied for the investigation of these layers [26, 55]. It has been used, mainly, to confirm the formation of silicon carbide in the implanted layer and, to obtain the new information about the layer structure as well. In some papers, the dependence of both shift of the wave length of a minimum of IR transmission peak and the change of its half width versus the annealing temperature are used for interpretation of IR transmission spectra. In our opinion, it is necessary to investigate such important characteristics as a change of an area under a curve of IR transmission spectrum band and a change of peak amplitude at 800 cm^{-1} versus the annealing temperature which contains very valuable information about the structure changes in an implanted layer. If the thickness of ion-synthesized film is comparable or smaller than the wavelength of incident electromagnetic radiation, under certain geometric conditions of the experiment, one can observe not only the transverse optical oscillations of atoms (TO-phonons), as well as longitudinal optical lattice oscillations (LO-phonons) [3]. The detection of LO-phonon peak of SiC and its change after film annealing give additional information on the crystallization processes. It is necessary to carry out the circumstantial investigations devoted to an analysis in detail of change in a wider temperature interval of an half width of IR transmission peak which characterizes the degree of structure order of an ion implanted layer.

In the majority of analogous studies, the post-implantation isochronous annealing of samples was carried out at temperatures from 400 up to 1200°C [11, 14, 2, 27]. However, in several studies [28, 47] the temperature range was extended and, the ion implanted layers had been annealed at temperatures 1300 and 1405°C. We believe also, that the temperature interval 400-1200°C is not sufficient for a annealing of disordered layer and completion of crystallization processes. A more detailed investigation of processes at temperatures ranged from 20 up to 1405°C permits to observe a number interesting effects taking place in an implanted layer.

The authors of papers [11, 2, 5, 55] declare about a significant diffusion of carbon and, contrariwise, that is negatived in works [13, 26, 27]. The authors of papers [8, 4, 47, 34] show that a layer has the electron conduction after annealing. The data of [28] give of evidence about the p-type conduction.

In a number of studies [6, 2, 5] a synthesis of SiC on a (100) oriented silicon substrate is considered as preferable, but at the same time in papers [13] the orientation (111) of substrate is declared as a most suitable. [4] investigated the optical and photoelectric properties of the SiC-Si structure, formed by implantation into (100), (110) and (111) oriented n- and p-type silicon of ^{12}C ions with energies of 40 and 70 keV, and doses of 4.3×10^{17} and 5×10^{17} cm^{-2}. Analysis of the IR absorption spectra of silicon layers implanted by carbon ions with energy of 70 keV allowed finding a significant dependence of crystallinity of the SiC layer on the orientation of the substrate after annealing at temperatures of 700–900°C. Although the tetrahedral Si–C-bonds more intense formed at an orientation of the substrate (100), annealing at 1100°C all evens out differences in the absorption spectra for all three substrate orientations (100), (110) and (111). The photovoltage photoresponse was obtained in all implanted structures. Investigation of current-voltage characteristics showed improvement in the rectification effect of the structure after annealing. The possibility of creating of β-SiC—Si heterostructures by ion implantation technique (band gap of 2.39 eV and 1.11 eV, respectively) was shown.

Frangis et al. [17, 18] formed β-SiC in silicon by high-temperature implantation (850–950°C) of carbon ions with energies 200 keV and doses ranged within $(0.2\text{–}1)\times10^{18}$ cm^{-2}. Implantation was carried out into silicon wafers of orientation (100) and (111). In both cases, β-SiC was formed with the same orientation as the matrix. It also reported that implantation at a lower temperature (500°C), but at higher energy (300 keV) leads to the formation of good quality β-SiC.

Aleksandrov et al. [6] carried out the synthesis of single-crystal SiC layer with one-step technique of high current ion implantation of carbon atoms into silicon substrates with orientations (001) and (111). Single crystal layer of SiC, which contains a small number of twins, was synthesized by the implantation of carbon ions with dose of 6×10^{17} cm^{-2} into (001) oriented silicon wafer using a focused ion beam with current density of 300 A/cm^2. When the ion current density was 150 A/cm^2, a single crystal SiC layer with a high concentration of twins was formed at the interface with the substrate Si. On top of this layer is formed a layer of polycrystalline SiC. When the implantation of carbon ions was carried out into (111) oriented silicon, single crystal β-SiC layer is not formed even when implanted into substrate heated up to a temperature of 850°C. Polycrystalline SiC layer at the surface and single-crystal SiC layer with a high density of twins near the interface with the crystal Si matrix, are formed.

This chapter presents the study of silicon carbide and carbon layers on silicon synthesized by ion beam techniques. The investigations of silicon layers implanted by carbon ions with energy 40 keV and dose 3.56×10^{17} cm^{-2} after annealing over a wide temperature range from 20 up to 1400°C using the special IR analysis are described. The features of change of the SiC-peaks in the spectra of the infrared transmission due to the influence of the Gaussian profile of the distribution of carbon in silicon are shown. Experiments to observe the longitudinal optical oscillations (LO-phonons) associated with the silicon carbide were carried out. A type of a conduction of synthesized silicon carbide was studied. Definite information from a shape analysis of the IR transmission curve was obtained. A particular attention was attracted on some problems which were disputable in previous investigations. IR studies of high-temperature instability of homogeneous layers of silicon carbide on (100) and (111) oriented silicon substrates synthesized by multiple implantation of carbon ions with energies E = 40, 20, 10, 5 and 3 keV, are described. By IR spectroscopy, Auger electron spectroscopy and X-ray reflectometry the composition and the processes of structural adjustment of the layer during annealing are analyzed. By ion-beam sputtering and magnetron sputtering techniques the $SiC_{0.8}$ and C films on the silicon wafers were deposited. Characteristics of the films by X-ray reflectometry are analyzed.

2. EXPERIMENTAL

Carbon implantation was carried out under completely oil-free conditions using elaborated accelerator. A vacuum in the implantation chamber is created by help of the ceolite vacuum pumps (6.5×10^{-2} Pa) and titan magnet discharge pumps (1.3×10^{-4} Pa). These pumps permit to except completely the organic compounds in volume which could be to contaminate the surface of the implanting samples. Gas (dioxide of carbon) has been used to obtain the single-charged ions of $^{12}C^{+}$. The implantation dose was determined by integrating of the beam current registrated on the target with suppression of a secondary emission. In order to prevent a sample heating during implantation, the ion current density was kept at a level less than 3 mkA/cm^2. The temperature of the target during the implantation is controlled by a thermocouple and, it not exceeds 20–25°C. The implantation of carbon ions was carried out into single-crystal (100) and (111) oriented silicon wafers of sizes $7\times12\times0.4$ mm^3 with an electrical resistivity 4–5 and 10 Ohm cm, respectively. After cleaning and removing the native surface oxide in a chemical etch, the samples were mounted in the target chamber of the implanter.

A set of these silicon wafers were implanted by $^{12}C^{+}$ ions with energy 40 keV and dose 3.56×10^{17} cm^{-2} at room temperature. To observe the longitudinal optical oscillations (LO-phonons) of atoms in synthesized film, a rotating shaft was incorporated into work chamber of infrared spectrometer. A sample holder is attached on this shaft. This system permits to make the IR transmission measurements of an ion implanted layer versus an angle of incidence of electromagnetic radiation on sample surface over the range 0-360° with step of 5°. However, in practice, at measuring of spectra we change the angle of incidence from 0 up to ±75°. It was observed no differences in transmission spectra measured from samples sloped to the radiation at a rotation of shaft both clockwise and anti-clockwise from the normal. Isochronous annealing of ion implanted samples was carried out in vacuum over the temperature range 200-1400^{0}C with steps of 50-200°C. The annealing was carried out in a low-inertia economical vacuum furnace especially elaborated and created for these purposes. It was carried out in conditions of completely oil-free pumping-out at a residual pressure ~ 1.3×10^{-4} Pa. The temperature was controlled by a help of tungsten-rhenium thermocouple.

To obtain a rectangular profile of the distribution of carbon atoms in the silicon, implantation of carbon ions of different energies and doses into second set of single-crystal silicon wafers of n- and p-type of conductivity was carried out sequentially in the following order: 1) E = 40 keV, D = 2.80×10^{17} cm^{-2}, 2) 20 keV and 0.96×10^{17} cm^{-2}, 3) 10 keV and 0.495×10^{17} cm^{-2}, 4) 5 keV and 0.165×10^{17} cm^{-2}, 5) 3 keV and 0.115×10^{17} cm^{-2}. The ratio of the concentrations of carbon and silicon atoms in the depth was about N_C/N_{Si} = 0.7. Post implantation annealing of the samples was performed in a vacuum in the temperature range 200-1200°C for 30 min with a step of 200°C. Then, the SiC films were subjected to prolonged isothermal annealing at the temperature of 1200°C for several hours in an atmosphere of inert gas (Ar) and, after specific time intervals infrared transmission spectra were recorded. The IR transmission spectra were recorded in differential regime on double-beam infrared spectrometer (400–5000 cm^{-1}). The spectra both at perpendicular incidence of infrared rays on the sample surface and at an angle of 73° with respect to the normal to the sample surface were measured. The composition of the layers was examined by Auger electron spectroscopy. The parameters were as follows: incident electron beam of diameter 1 μm, energy 10 keV, angle of incidence 45°, diameter of scanning region 300 μm, vacuum 1.33 $\times 10^{-8}$ Pa, angle of Ar^{+} beam incidence 45°. Parameters of films were investigated using the X-ray reflectometry at small glancing angles by recording the angular dependence of the reflection coefficient for two spectral X-ray lines CuK_{α} (0.154 nm) and CuK_{β} (0.139 nm) at the facility "ComplexXRay C6". Selection of spectral lines CuK_{α} and CuK_{β} from polychromatic spectrum was carried out using thin semi-transparent and thick untransparent monochromators, respectively, made from the pyrolytic graphite with a mosaic angle of 0.5°.

SiC films on the silicon substrates (25°C) were also synthesized by ion-beam sputtering of a two-component target of graphite and silicon. The C films on the silicon substrate (75°C) by magnetron sputtering were synthesized. Parameters of SiC and C films on Si substrates were determined using the X-ray reflectometry.

3. RESULTS AND DISCUSSION

3.1. A Model Of An Carbon Implanted Silicon Layer And The Mechanism Of The Low Temperature Formation Of Si- And Sic Crystallites

In Fig.1 the calculated profile of carbon atom distribution in Si constructed basing on data R_p and $\Delta R_p$from [21] are presented. The Gaussian profile (Fig.1, curve 1) was calculated for the implantation of carbon ions with energy 40 keV and dose 3.534×10^{17} cm^{-2}, when the carbon concentration in the distribution peak is equal to stoichiometric composition of SiC, i.e. N_C/N_{Si} = 1, where N_C/N_{Si} is the ratio of the concentrations of C and Si atoms. The curve 2 in Fig.1 shows the calculated profile, corresponding to dose D = 3.56×10^{17} cm^{-2} of carbon ions used in this investigation.

3.1.1. Lo-Phonons and Their Applications To Analysis Of An Implanted Layer

As it is well known [45, 66], during interaction of electromagnetic waves with an infinite crystal lattice, the transversal optical oscillations (TO-phonons) of atoms are excited. In overwhelming majority of previous investigations, the synthesis of silicon carbide was identified by help of spectra of transversal optical phonons. No attempts was made to detect the longitudinal optical oscillations (LO-phonons) of atoms of lattice for a Gaussian concentration profiles of carbon in spite of that majority of studies in the field of ion synthesis of silicon carbide was carried out using these profiles. [3] found that an absorption at wave number 980 cm^{-1} is observed, if the angle of incidence of irradiation on the sample surface deviate from perpendicularity. Implantation of carbon ions with energies of 24 and 40 keV and a dose of 4.3×10^{17} cm^{-2} carried out at room temperature into a (111) oriented Si plate of p-type conductivity. Annealing was performed in a vacuum at temperatures of 900 and 1100°C for 30 minutes. The presence in the transmission spectrum of bands associated with the LO-and TO-phonons made possible to calculate such parameters of SiC, as high frequency, ε_∞, and the low-frequency dielectric constant, ε_0, attenuation coefficient of the phonons, the effective charge e^*/e and the force constant ρ. So, the detection of LO-phonons may be used to obtain additional information about a structure of ion implanted layer. Morever, the frequency values of both the transversal- and longitudinal oscillations permit to determine the parameters of efficient charge which is a quantitative criterion of a compound polarity. The efficient charge value permits to calculate a mobility of free charge carriers.

If a thickness of ion synthesized film is less or comparable with a wave length of the electromagnetic radiation incident on film's surface, the limitations related with the condition of infinity of crystal lattice are lifted. As a result, one can to observe the longitudinal optical oscillations of lattice atoms at definite geometrical conditions of experiments [66]. In this relation, the special experiments to observe these phonons were carried out. For this purpose, the IR transmisssion spectra versus an angle of incidence of electromagnetic radiation on sample surface with step of 5° were measured.

In Figs.2 and 3 the IR transmission spectra of both (100)- and (111) oriented silicon samples implanted by carbon ions with energy 40 kev and dose 3.56×10^{17} cm^{-2}, after isochronous annealing over the temperature range 200-1400°C, are presented. The spectra at both perpendicular incidence of infrared rays on the sample surface and at an angle of 73° with respect to the normal to the surface were measured. In Fig.4 the wave number values in maximum of IR transmission versus the annealing temperature are presented. The curves on this figure were constructed using the experimental data presented on Figs.2 and 3. The curves for TO-phonons were constructed basing on the infrared transmission spectra measured by using of perpendicular incidence of the infrared rays on the sample surface.

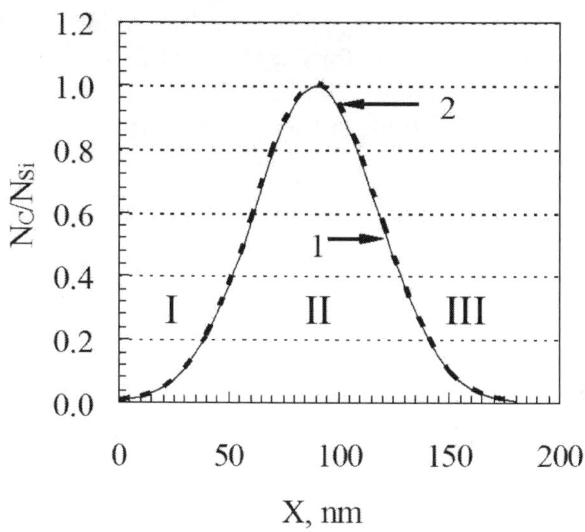

Figure 1. The calculated profiles of distribution of carbon atoms in Si constructed basing on data of R_p and ΔR_p from [21]. The energy of carbon ions is 40 keV, dose 3.534×10^{17} $см^{-2}$ (curve 1) and $3.56\times10^{17}см^{-2}$ (curve 2).

Figs. 2 and 3 show the appearance of an IR absorption peak at 965-970 cm^{-1} in the spectra from samples inclined to IR irradiation at an angle of 73° with respect to the normal to the sample surface. This absorption peak begins to be appeared after annealing at 1000°C for both types of substrate orientation together with the main peak at 797-800 cm^{-1} which corresponds to the transversal optical atomic oscillation of SiC.

Basing on the values of wavelength of this peak, its amplitude and synchronous modification together with the peak for transversal optical oscillation of SiC during annealing as well, the peak at 965-970 cm^{-1} was associated with longitudinal optical phonons of silicon carbide. The temperature of LO-phonon appearance is about 300°C higher, than that for TO-phonons of SiC (700°C). It may be caused by smallness of the LO-phonon peak amplitude and, in consequence of this, by difficulties of their registration.

For (100) oriented substrates, the increase of the annealing temperature over the range 1000–1350°C leads a linearly increase of LO-phonon wavelength at minimum of amplitude from 930 to 965 $cm^{-1}$and, then wavelength increasing is saturated. As for (111) oriented substrates, unlinearly increase of LO-phonon wavelength in the range 955–970 cm^{-1} up to 1400°C is observed. Thus, the formation of SiC crystallites, i.e. the intensive formation of Si-C tetrahedral bonds of necessary length and bond angles, in the case of (111) oriented substrate is not completed up to the silicon melting point, as for (100) oriented substrate that is completed at 1350°C. The difference between the LO-phonon curves behaviour for (100) and (111) oriented substrates indicates on the differences in the crystallization mecanism. One can see an influence of silicon substrate orientation on SiC crystallization in the implanted layer from the TO-phonon curve also (Fig.4).

In some investigations no changes in the IR transmission spectra after annealing at 700°C [8], 850°C [14], 875°C [11], 900°C [26], 1100°C [2] were observed. That was explained by finishing of the β-SiC formation process. Really, that may be correct, if we are based on the analysis of TO-phonon curve only. However, as is seen from Fig.4, the TO-phonon curves show the saturated absorption and give no addititional information over the temperature range of 900–1400°C, as the LO-phonon curves undergo the substantial changes at these temperatures indicating on the structural changes in the ion implanted layer. Thus, one can conclude that the observation and measurement of LO-phonon peak are important for an analysis of crystallization process.

As it is known [38], the TO- and LO-phonons frequences are bounded with the equation:

$$\frac{\nu_{LO}}{\nu_{TO}} = \left(\frac{\varepsilon_0}{\varepsilon_\infty}\right)^{1/2} \qquad (1)$$

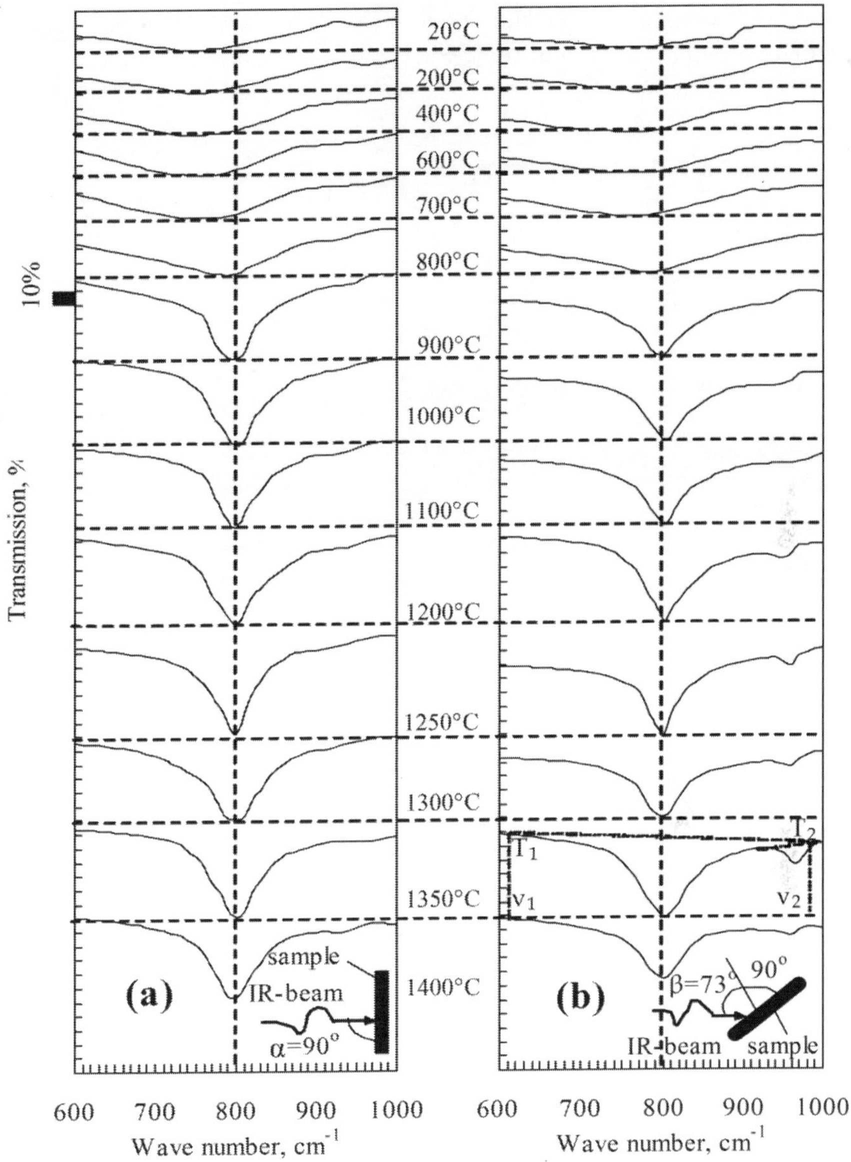

Figure 2. The IR transmission spectra of (100) oriented Si samples implanted by $^{+}C^{12}$ ions (E = 40 kev, D = 3.56×10^{17} cm^{-2}, after isochronous annealing over the temperature range 200-1400°C: a) α = 90°; b) β = 73°.

Where ε_0 and ε_∞ are the low-frequency and the high-frequency dielectric constants, respectively. The effective charge e*/e is determined from the equation [54]:

$$\frac{e^*}{e}=\left(\frac{\varepsilon_0-\varepsilon_\infty}{4\pi}\right)^{1/2}\left(\frac{M_n}{N}\right)^{1/2}\left(\frac{3\omega_0}{\varepsilon_\infty+2}\right)$$

$$\left(\frac{\pi(\varepsilon_0-\varepsilon_\infty)M_n\nu_{TO}^2}{Ne^2}\right)^{1/2}\left(\frac{3}{\varepsilon_\infty+2}\right)$$

(2)

where ω_0 is the resonance frequency, $N = 4.84\times10^{22}$ cm^{-3} is the concentration of ion pairs, $M_n = (M_+M_-)/(M_++M_-) = 1.396\times10^{-23}$ g is the reduced mass of ion pair, $\nu_{TO} = 2.395\times10^{-13}$ cm^{-1} is the frequency of TO-phonon irradiation.

The dimensionless parameter, ρ, which is proportional to the absorption quantity, is determined from the equation:

$$\rho=\frac{\varepsilon_0-\varepsilon_\infty}{4\pi}$$

(3)

The values of ε_0, e*/e and ρ determined from the equations (1)—(3) are equal to 9.82, 0.89, and 0.25, respectively. The value of ε_∞ has been chosen to be equal to 6.7, because according to [45] the sufficiently great dispersion of this value leads an insignificant changes. So, both the detection and the measuring of LO-phonons have been permited to determine some characteristics of synthesized film.

3.1.2. The Ir Transmission Analysis Of Ion-Implanted Layer On (100) And (111) Oriented Silicon

In Figs.5a and b, the values of amplitude and halfwidth (FWHM) of the IR transmission peak, respectively, versus an incidence angle of the infrared radiation on the carbon implanted (100) oriented silicon samples, are presented. These data were obtained after isochronous annealing of samples over the range 200-1200°C for 30 min with the step of 200°C. Beginning from an angle α = 50° (Fig.5a),almost linear decreasing of TO-phonon peak amplitude and simultaneous increasing of LO-phonon peak amplitude are observed. The changes of LO- and TO-phonons peak amplitudes are correlated with one another.

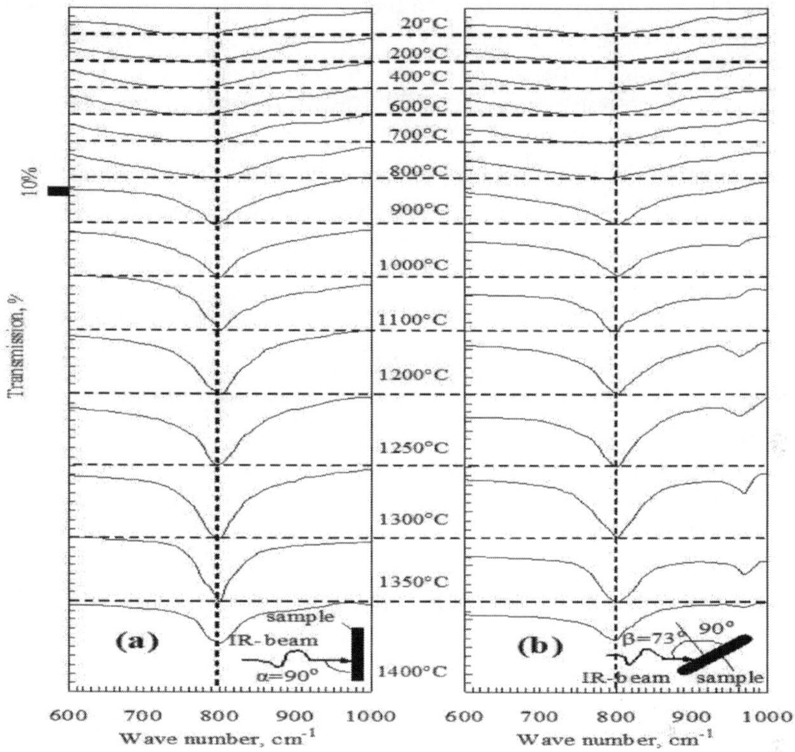

Figure 3. The IR transmission spectra of (111) oriented Si samples implanted by $^{+}C^{12}$ ions (E = 40 kev, D = 3.56×10^{17} cm^{-2}), after isochronous annealing over the temperature range 200–1400°C: a) α = 90°; b) β = 73°.

The halfwidth (FWHM) changes of the peak have a more complicated dependence from the incidence angle of the infrared radiation on the sample surface (Fig.5b). This dependence has no correlation with the data in Fig.5a. The half-width of the peak is usually dependent on the quality of the crystal structure of the film and should not depend on the angle of incidence of IR radiation. Apparently, the decreasing of halfwidth of the TO-phonon peak can be explained by the presence of non-tetrahedral Si-C-bonds of a certain type, which are oriented in the space of the film in such a way that with increasing angle α above 35° they cease to absorb infrared radiation at frequencies near 800 cm^{-1}.

This is equivalent to the effect of decay of these bonds, since it leads to a decrease in the amplitude and the narrowing of the TO-phonon peak, but can not testify about improving the structure of the layer. This is also accompanied by the appearance of deformed LO-phonon peak in the frequency range near 950 cm^{-1}. With the increase of the angle α up to 73°, the narrowing of LO-phonon peak and an increase in its amplitude are taken place. The interpretation of these results requires further investigation.

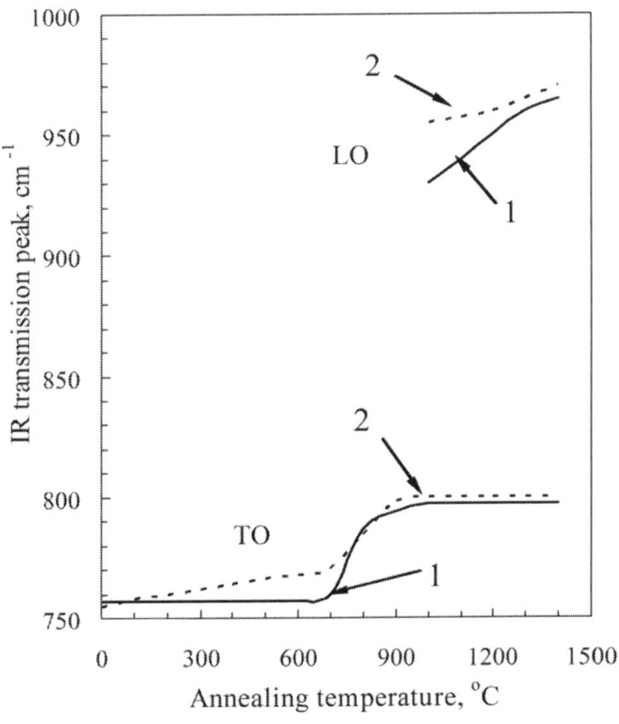

Figure 4. Wave numbers of IR transmission minimum for TO- and LO-phonon peaks of SiC versus an annealing temperature for carbon implanted silicon layers on Si substrates (E = 40 kev, D = 3.56×10^{17} cm^{-2}): 1 - Si(000); 2 - Si(111).

In previous IR investigations of implanted by $^{12}C^{+}$ ions silicon layers, a shift in frequency of an absorption maximum versus the annealing temperature, and also the changes of half-width and amplitude of peak were observed. Analyzing the IR transmission spectra presented in

Figs.2 and 3, one can see not only the changes of these parameters. A base-line to each spectrum (Fig.2, 1350°C) was drawn. As is seen, the areas of obtained figures are changed, too (Fig.6). In our opinion, the area under the IR transmission curve is associated with the number of absorbing objects in the ion-implanted layer and better shows the transformation of these objects during isochronal annealing. As the object may be not only the crystallites of SiC, but also the another types of infrared active compounds of carbon atoms with carbon or silicon atoms, and silicon atoms one with another, which one can unify under one common appellation - clusters.

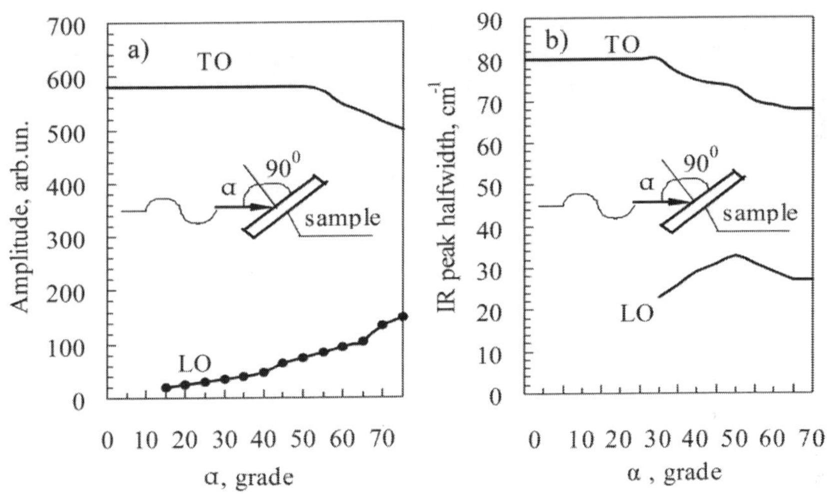

Figure 5. Amplitude (a) and halfwidth (FWHM) (b) of TO- and LO-phonons peaks of SiC of the IR transmission versus the incidence angle of IR radiation on the surface of the carbon implanted Si.

Basing on the mentioned above, the IR transmission spectra (Figs.2 and 3) were analyzed in detail accordingly to all listed points. In Fig.6 an area of the IR transmission peak (see Fig.2) associated with TO-phonons of SiC obtained both at perpendicular incidence of infrared rays on the sample surface (curve 1) and at an angle of 73° with respect to the normal to the sample surface (curve 2), versus the annealing temperature are presented. Area values can be determined by direct measurement or by using the expression (Fig.2, 1350°C):

$$A = \frac{1}{2}(T_1 + T_2)(\nu_2 - \nu_1) - \int \tau(\nu)d\nu \approx \frac{1}{2}(T_1 + T_2)(\nu_2 - \nu_1) - \sum \tau(\nu)\delta\nu \quad (4)$$

where A – total absorption (or transmission) in relative units in the wave number range $\nu_1<\nu<\nu_2$, $\tau(\nu)$ – transmission at frequency ν, T_1 and T_2 – the values of IR transmission at wave numbers ν_1 and ν_2, respectively, $\delta\nu$ – step of measurements, equal to 2.5 or 5 cm^{-1}. The areas corresponded to LO-phonons have been not measured due to of their infinitesimal. Further the data obtained for a perpendicular incidence of IR-rays on sample surface will be discussed, as an analysis of the curve 2 is difficult due to the absence of the reflection data. As is seen from Fig. 6 (curve 1), the four peaks at 600, 1000, 1200 and 1350°C are evidently observed, which, seemingly, are related with four physical processes occuring in ion implanted layer in four temperature ranges. The same maxima are observed for curve 2 in approximately the same temperature ranges. On this curve a fifth maximum at 200°C is also observed.

It is necessary to note, that in most previous investigations, mainly, the informations corresponding to the peak in range 900-1000°C are obtained using the dependences of halfwidth and a shift in frequency of an absorption maximum from the annealing temperature. The physical processes corresponded to the peaks of both 600 and 1200-1400°C have not been studied in detail, partly, due to the hard access of the temperature range of 1200-1400°C and, partly, due to weak defined processes at 600°C. It follows from the Fig.6 that the change of area of SiC-peak takes place over the whole temperature range from 20°C up to 1400°C.

In Fig.7 the values of IR transmission amplitude for TO-phonons of SiC at wavenumber 800 cm^{-1} and for LO-phonons of SiC versus an annealing temperature for spectra presented in Figs.2 and 3 (curves 1, 1' – for the perpendicular incidence of IR rays on sample, the curves 2, 2', 3, 3' – for an angle of 73°), are shown. When constructing these dependences we have believed that the IR transmission amplitude at 800 cm^{-1} is proportional to the concentration of tetrahedral oriented Si–C-bonds of atoms incorporated into crystallites of SiC. All factors which can affect on a broadening of peak corresponding to an infinitely thin ideal film of SiC have been neglected. This approximation is to some extent, may distort the true picture of the physical phenomena occurring in ion-implanted layer. However, this assumption is very important in a qualitative sense, since it allows understanding the general course of the process and separating a region of the IR transmission peak [30] due to the contribution of crystallites of SiC in the area value, from a region due to the optically active clusters. As seen in Fig.7, the overall shape of the curves 1 and 2 for the TO-phonons is about the same and has a number of features, in particular, at 1300°C, which is also found on the curve 3 for the LO-phonons. The changes of amplitude take place over the all temperature range from 20 up to 1400°C. There are differences in the crystallization processes of carbon implanted silicon layers for the orientation of the substrate (100) and (111).

In Fig.8 the half-width of the TO-phonon peak of SiC at perpendicular incidence of IR radiation on sample surface for spectra shown in Figs.2 (curve 1 – for substrate Si(100)) and 3 (curve 1' – for substrate Si(111)), versus an annealing temperature, is presented. The analogous dependences have been presented in previous papers [8, 47, 26]. However, the temperature range was narrower and, therefore, the maxima at 1200 and 1350°C were not found.

To explain these effects let us consider the ideal case again. We assume that the broadening of the absorption band due to different processes is absent and, therefore each frequency in the transmission spectrum corresponds to one or other bond between the atoms of the implanted layer. On this basis, we can see that, immediately after the implantation the contour of the transmission curve covers a wide range of frequencies, i. e. in ion-implanted layer there are many different bonds that absorb at different frequencies. If one attributes the frequency of 800 cm^{-1} to the tetrahedral oriented Si-C-bond of length of 0.194 nm (bond characteristic of the silicon carbide), so in the implanted layer there are the systems with the bond lengths of both larger and smaller than this. In general, the presence of different bond lengths between the atoms of the ion-implanted layer is completely natural because atoms can stop at different distances from each other in the process of implantation.

In our case, the most interesting bonds are the single-, double- and triple silicon-silicon (Si–Si, Si=Si, Si≡Si), silicon-carbon (Si–C, Si=C, Si≡C) and carbon-carbon (C–C, C=C, C≡C) bonds presented inFig.9. Simple covalent bond C–C, formed by the overlap of two sp3-hybrid electron clouds along the line connecting the centers of atoms, is σ-bond. One of the electron pairs in the double C=C bond forms σ-bond, and the second bond is formed by p-electrons with the clouds in the form of "eight", which overlapping, form a π-bond. Triple C≡C bond is a combination of one σ-bond and two π-bonds [22]. The lengths of single bonds are shown in proportion to ones which are characteristic for these bonds in a tetrahedral orientation, although they can have various values in the ion implanted layer. And in the case of double and triple bonds they may be either higher or lower than the values given. The length of a single bond of the same type of atoms was taken equal to twice the covalent radius of atoms, and the length of the Si–C-bond was taken as half the sum of double the values of covalent radii of Si- and C without correction for their ionicity. The lengths of double and triple bonds were taken at 0.021 and 0.034 nm shorter than a single bond, respectively. Triple bond between two atoms can be represented by two tetrahedra sharing a common face, and for the double bond - by two tetrahedra sharing a common edge [44].

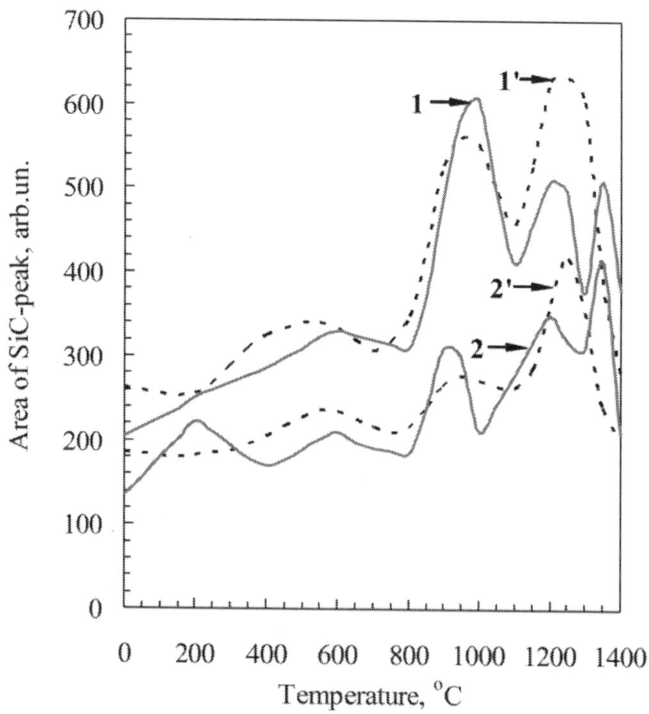

Figure 6. An area of the IR transmission peak for TO-phonons of SiC at perpendicular incidence of the radiation on the sample surface (curves 1, 1') and at 73° from a normal (curves 2, 2') for spectra presented in Figs.2 (curves 1, 2 - substrate Si(100)) and 3 (curves 1', 2' - substrate Si(111)), versus an annealing temperature.

In the process of implantation of carbon into silicon vast majority of the covalent bonds of the substrate, starting with the amorphization threshold, are not covalent, because of violation of bond lengths and angles between them. However, among the formed Si–C-bonds there are tetrahedral bonds, the distances and angles between atoms of which correspond exactly to the crystallites of silicon carbide. This is confirmed by the presence of absorption at 800 cm^{-1} and by the results of the authors [26] who identified by electron diffraction the presence of silicon carbide crystallites immediately after the implantation of carbon into silicon.

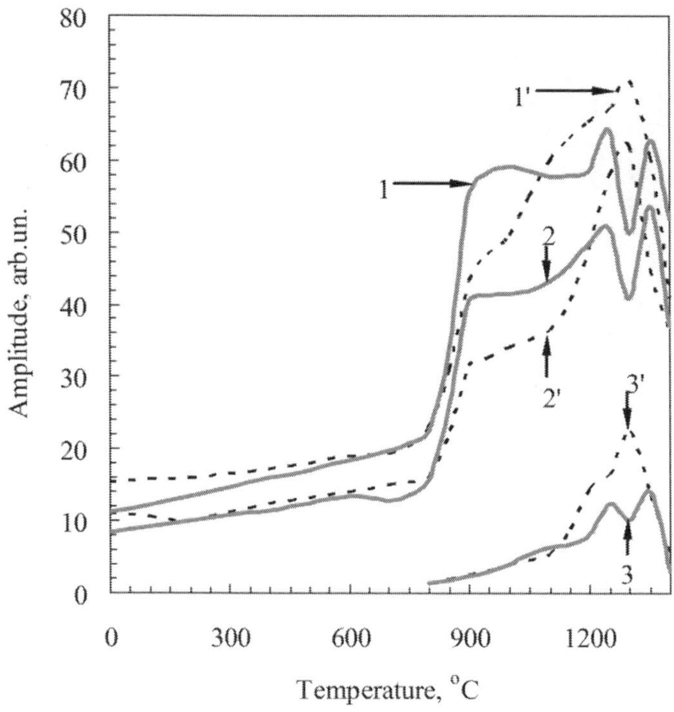

Figure 7. The IR transmission amplitude values for TO-phonons of SiC (curves 1, 1', 2, 2') at 800 cm^{-1} and for LO-phonons of SiC (curves 3, 3') at perpendicular incidence of the radiation on the sample surface (curves 1, 1') and at 73° from a normal (curves 2, 2', 3, 3') to surface for spectra presented in Figs.2(curves 1, 2, 3 - substrate Si(100)) and 3 (curves 1', 2', 3' - substrate Si(111)), versus an annealing temperature.

We believe that the ion-implanted layer consists mainly of various combinations of the nine types of bonds, shown in Fig. 9. Moreover, it is possible the presence in the implanted layer of the single elongated bonds, sesqui, free ("dangling") and hybridized bonds, as well as resonances and other higher order interactions as well. By making these assumptions, we proceed not only from the contour of the spectrum of infrared transmission, covering a wide range of frequencies. We are basing also on the ability of carbon and silicon atoms to form besides single bonds also double and triple bonds [29, 44,64, 46]. In paper [27], an aggregate of carbon atoms was named as a cluster. In this paper, as a cluster we have in mind all the nine types of bonds and combinations thereof, from which are formed during annealing a three-dimensional clusters and crystallites of Si and SiC.

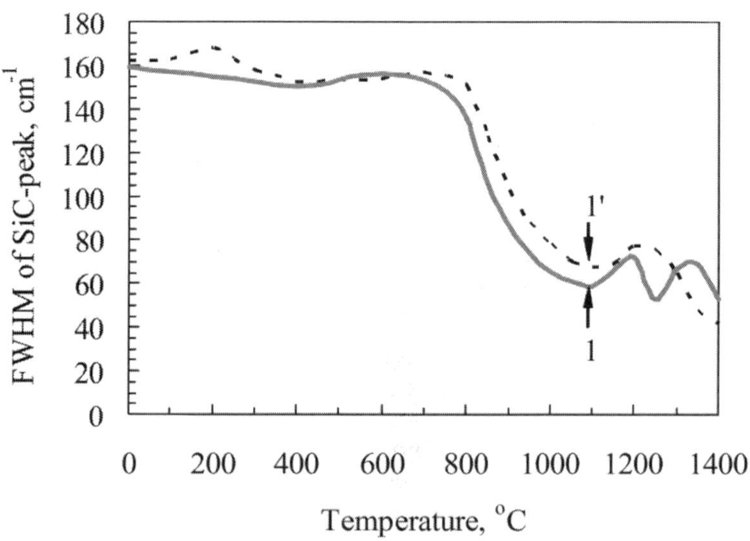

Figure 8. Half-width of the TO-phonon peak of SiC at perpendicular incidence of IR radiation on sample surface for spectra presented in Figs.2 (curve 1 – substrate Si(100)) and 3 (curve 1' – substrate Si(111)), versus an annealing temperature.

Table 1 presents the values of the binding energy for all nine types of bonds, shown in Figure 9. The sum of the energies of two and three single C–C bonds are equal to 688 and 1032 kJ mole^{-1}, respectively, such that by 73 and 220 kJ mole^{-1} higher than energy values of the C=C and C≡C bonds listed in Table 1. This suggests that the structures of the C=C and C≡C bonds for one bond has less energy than the structures with a single C–C bonds.

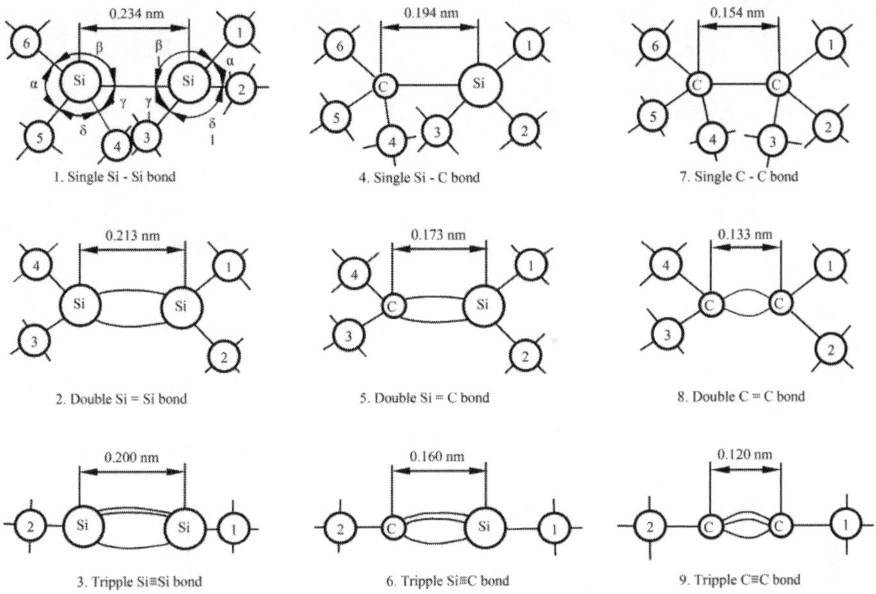

Figure 9. Various types of bonds between silicon, carbon atoms or their combination.

Table 1. The values of binding energy for nine types of bond.

Types of bond	Binding energy, kJ mole^{-1}	Types of bond	Binding energy, kJ•mole^{-1}	Types of bond	Binding energy, kJ•mole^{-1}
Si–Si	187	Si–C	290	C–C	344
Si=Si	<374	Si=C	<580	C=C	615
Si≡Si	<561	Si≡C	<870	C≡C	812

By analogy, reasonable to assume that clusters Si=Si, Si≡Si, Si=C, Si≡C per bond also has less energy than clusters with single Si–Si and Si–C bonds. Consequently, the energy of double and triple bonds Si=C, Si=Si, Si≡C and Si≡Si must be less than the sum of the energies of two or three single Si–Si and Si–C bonds, as is shown in Table 1. It is evident that the most strongly bonds are the carbon-carbon and then carbon-silicon and silicon-silicon clusters.

Let us consider the special features of the curves on Figs.6–8 basing on the assumptions presented above.

The temperature range 20-600°C

It is known [65] that a typical recrystallization temperature of amorphous silicon lies in the temperature range 500–600°C. When the dose of the implanted carbon ions is much higher than the amorphization threshold and, the implanted atoms combining with the silicon atoms can form the inclusions of new compounds of considerable volume, the crystallization of silicon, in the case of a Gaussian distribution profile of implanted atoms, starts at the surface and the interface "disturbed layer – substrate" and goes in the direction to the maximum of the carbon distribution with increasing annealing temperature. The silicon crystallites are formed in regions where the concentration of silicon atoms exceeds the concentration of carbon.

The values of concentration ratio N_C/N_{Si} (Fig.1) are small on the edges of distribution and, a significant number of silicon atoms falls at each implanted carbon atom. In this temperature range, such combinations of clusters in ion-implanted layer are decaying which consist mainly of bonds of Si–Si, Si=Si and elongated Si–C, as they have the lowest energy dissociation among the types of bonds listed above. As decay of the clusters, we mean such regrouping of the atoms in system and the change the lengths of chemical bonds and angles between them, which lead to the most energetically favorable state of system. In such state there is a system of atoms with tetrahedrally oriented bonds, which are the most stable and durable. All other types of bonds and their geometrical arrangements are energy unprofitable; they are insufficiently stable and can decay during annealing.

As it is seen from curves 1 and 1' in the Figs. 6 and 7, the increasing of both an area of SiC-peak of IR transmission and its amplitude at 800 cm^{-1} takes place over the temperature range 20–600°C, i.e. the formation of SiC crystallites takes place at temperatures significantly less that it is necessary for the SiC formation by a thermal growth. The mechanism of this phenomenon is of interest.

In the system "implanted layer – substrate", the phonons are generated in the process of annealing. This system is a single entity and, it is reasonable to assume that between the implanted layer and the substrate there is a continuous interaction of phonons. Since the thickness of the substrate is much greater than the thickness of the implanted layer, then, with respect to the layer, the substrate can act as a huge reservoir of phonons, which able, due to its heat capacity, receive or deliver the phonons to the implanted layer. As the implanted layer is thermodynamically nonequilibrium system, the process of interaction of phonons with the atoms will go towards reducing the free energy of the system. During each collision of a phonon with cluster of the implanted layer, the phonon absorption will occur if the energy of the system will decrease.

Energy released during the decay of clusters, is transferred to the lattice, which can accumulate and transfer it to another cluster, followed by the formation of energetically favorable system, in this case, the crystallites of Si and SiC. It is also possible direct transfer of energy of the decaying Si–Si-cluster to other types of clusters. In this temperature range, mainly, weakly bound Si–Si-clusters can decay during the interaction with phonons. More energy requires for the formation of crystallites of Si and SiC. This energy is transfered to the reacting atoms by the lattice. The considered above mechanism of formation of crystallites of Si and SiC, the so-called over-barrier mechanism, when the interacting atoms overcome the energy barrier of height $E \cong E_n$, is shown in Fig.10. This mechanism of formation of tetrahedrally oriented bonds Si–Si and Si–C from an energy point of view is advantageous for the crystal lattice, since it thereby reduces its free energy.

All possible mechanisms of the tetrahedral oriented Si–Si- and Si–C-bonds formation should be energy profitable for the crystalline lattice to decrease free energy of system. The formation of crystallites of Si and SiC in the range 20–600°C occurs mainly due to the decay of clusters such as the longest Si–Si- and Si–C-bonds, but also partly due to the disintegration of other types of clusters. A number of studies have shown [5, 26], that immediately after the implantation, the presence of a minute quantity of SiC crystallites with hexagonal structure is observed in the ion implanted layer and, they are transformed into β-SiC at the temperature of 400°C and higher. It is impossible to identify the presence of Si crystallites due to their optical inactivity in this range of the infrared absorption. However, their presence is not in doubt, as much less energy is required to expend for their formation than for the formation of SiC crystallites at the implementation of over-barrier mechanism. As we showed earlier, in layers $SiC_{0.03}$ with a low carbon concentration the crystallites of Si increase their sizes from 2 to 3 nm in the range 20–700°C (Fig.11).

Peak area immediately after implantation is not zero, i. e. a part of the carbon atoms is included into composition of the optically active clusters (Fig.6). If we assume that, after annealing at 1000-1250°C almost all carbon atoms are optically active and optically inactive clusters broke up, we see that immediately after the implantation of carbon into the (100) and (111) oriented silicon at least 65% and 60% of carbon atoms were concentrated in optically inactive clusters, respectively, if the implantation was carried out by a dose sufficient to obtain the stoichiometric concentration (E = 40 keV, D = 3.56×10^{17} cm^{-2}). A number of carbon atoms is included into stable types of optically inactive clusters which stable up to melting point of silicon. It is known [44], that the optically inactive objects consist of the clusters and their chains that lie in one plane. The formation of clusters and chains of planar systems of nets may be due to energy considerations. For example, in [5] the formation of alternating layers of single crystal silicon with amorphous silicon precipitates enriched with carbon in the ion-implanted layer, attributed to the fact that the system in such a way reduces its free energy.

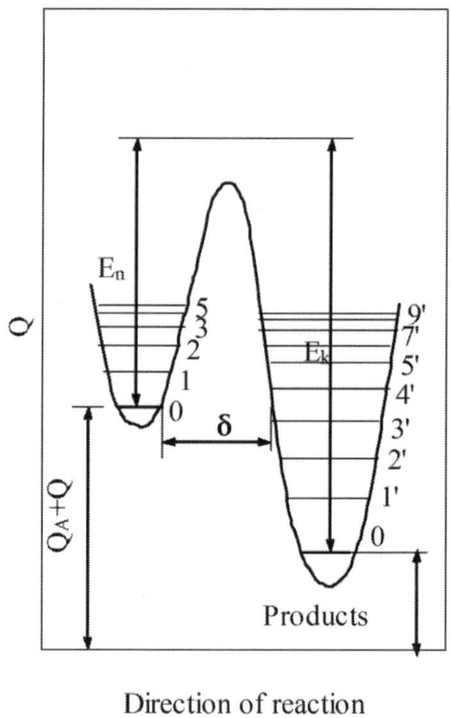

Figure 10. Illustration of the over-barrier mechanism of the formation of the tetrahedral oriented bonds

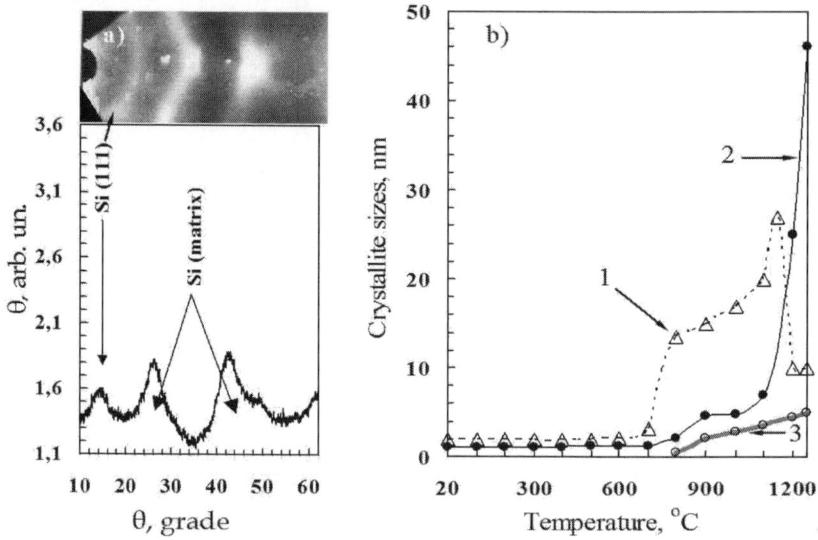

Figure 11. X-ray diffraction patterns of the $SiC_{0.12}$ layer after implantation (a) and average sizes of crystallites in the (111) plane after implantation and annealing (b): 1 – Si (for layer $SiC_{0.03}$), 2 – Si (for layer $SiC_{0.12}$), 3 – SiC (for layer $SiC_{0.12}$).

Spatial pattern of ion-implanted layer is difficult to model, since a Gaussian distribution profile of implanted atoms is characterized by the change by depth of the concentration of carbon atoms $N_C/N_{Si}$and, thus, the mechanism of physical processes from one layer to layer is changed. We can construct a flat infrared inactive net in the middle of layer where $N_C/N_{Si} = 1$. Flat optical inactive net consisting of C and Si atoms, linked by single, double and triple bonds, may also contain free ("dangling") bonds of the silicon and carbon atoms. These bonds may connect to atoms of the other flat net or on the association of the atoms which do not lie in one plane. The atoms which do not lie in one plane, can form an association of optically active clusters.

With increasing annealing temperature at first the decay of elongated single bonds at two atoms united by a triple bond, is taken place. These pairs of atoms inhibit the diffusion of atoms in the layer. Then, the energetically unfavorable single bonds of atoms are disintegrated in the planar nets of clusters. The subsequent increase in annealing temperature would lead to the disintegration of the planar nets forming a number free carbon and silicon atoms, as well as pairs of Si and C atoms, linked together by multiple bonds. Free carbon and silicon atoms can move on short distances and join to form the crystallite Si or SiC, which is profitable from an energy point of view and, as a result, the energy of system is decreased.

In addition to the planar nets of clusters, the ion-implanted layer may contain long chains of clusters, which are also optically inactive. The chains can be formed by alternating different types of bonds, and their degradation temperature may be different. There are also local clusters non-interacting with the surrounding atoms and consisting of three-, four- and more atoms linked together by double bonds. They are most stable clusters due to the full richness of their bonds. Perhaps these clusters are not disintegrated up to the melting point of layer.

In Fig.12 the IR transmission amplitude versus the annealing temperature for different frequencies is shown. It is evident that clusters absorbing at frequencies of 850 and 900 cm^{-1}, in the range 20–600°C did not disintegrated, as their amplitude remains unchanged. The position of minimum of IR transmission peak does not change and is located at 757 cm^{-1} (Fig.4). This indicates the dominant role of one type of clusters, which absorb at 757 cm^{-1}. The increase of the amplitude of the peak (Fig. 12, curve 3) indicates an increase in the concentration of these clusters with increasing annealing temperature. At the same time the concentration of clusters, which absorb at 700 cm^{-1} and correspond to the elongated single bond, is simultaneously increased. The energy of both the formation and decay of these bonds is a least one ($E = h\nu$), as they absorb the radiation of lowest frequencies among considered. It follows from Fig.7 (curve 1) and 12 (curve 1), that the bonds being very similar to tetrahedral bonds of SiC which absorbs at 800 cm^{-1}, are formed in the implanted layer. That follows also from decrease of peak halfwidth (Fig.8) in temperature range 20–400°C. Its increase in range 400–600°C may be associated with intensive restructuring of Si–Si bonds of amorphous silicon before recrystallization at the surface and near the substrate, where the concentration of carbon is low. The increase in peak area of the TO-phonon SiC in this range (Fig. 6, curves 1, 1 ') is caused by an increase in the number of tetrahedral bonds and close to them, which absorbs in the range 750-850 cm^{-1} (Fig.12, curves 1, 3, 5).

No significant differences between the properties of the films on the substrate orientation (100) and (111) in this temperature range were observed, except for the fact that the area of SiC-peak and amplitude at 800 cm^{-1} are slightly higher for the orientation (111), which indicates a higher number of tetrahedral bonds of SiC and a smaller number of optically inactive clusters.

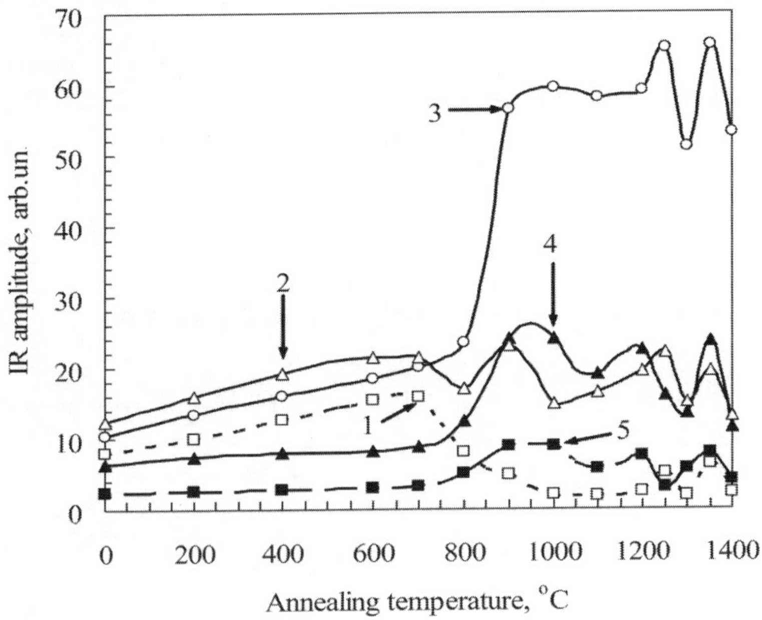

Figure 12. Amplitude of IR transmission for various wavenumber values versus an annealing temperature. (1 - □) 700 cm^{-1}, (2 - Δ) 750 cm^{-1}, (3 - ○) 800 cm^{-1}, (4 - ▲) 850 cm^{-1}, and (5 - ■) 900 cm^{-1}.

The temperature range 600-800°C

The area under IR spectrum curve is decreased in this temperature range (Fig.6) due to the decay of the optical active clusters absorbing at the frequencies close to 700 and 750 cm^{-1} (Fig.12). As shown in previous studies (Fig. 13d, e), this is due to the decay of elongated Si–C-bonds in the layers with a low concentration of carbon. An intensive process of disintegration of these bonds occurs in layers $SiC_{0.4}$and $SiC_{0.12}$ between the surface and the maximum of the carbon distribution. In addition of the decay of the optical active non-tetrahedral Si–C-bonds, seemingly, deformed Si-Si bonds are disintegrated, too.

The intensive rearrangement of clusters is characterized for this temperature range. As a result of multiple collisions the atoms of clusters, successively passing from the initial- through the intermediate states to the most energy favorable end position, form the tetrahedrally oriented bonds of Si- and SiC crystallites. A significant change of halfwidth of the Si–C-peak of IR spectrum begins for film on (100) oriented substrate (Fig.8). In case of (111) oriented silicon substrate the same is taken place some later. A certain increase of the concentration of clusters absorbing on the wavenumbers ranged from 850 to 900 cm^{-1} (Fig.12, curves 4, 5) is simultaneously taken place. The ordering of layer structure in region near the substrate is taken place, too.

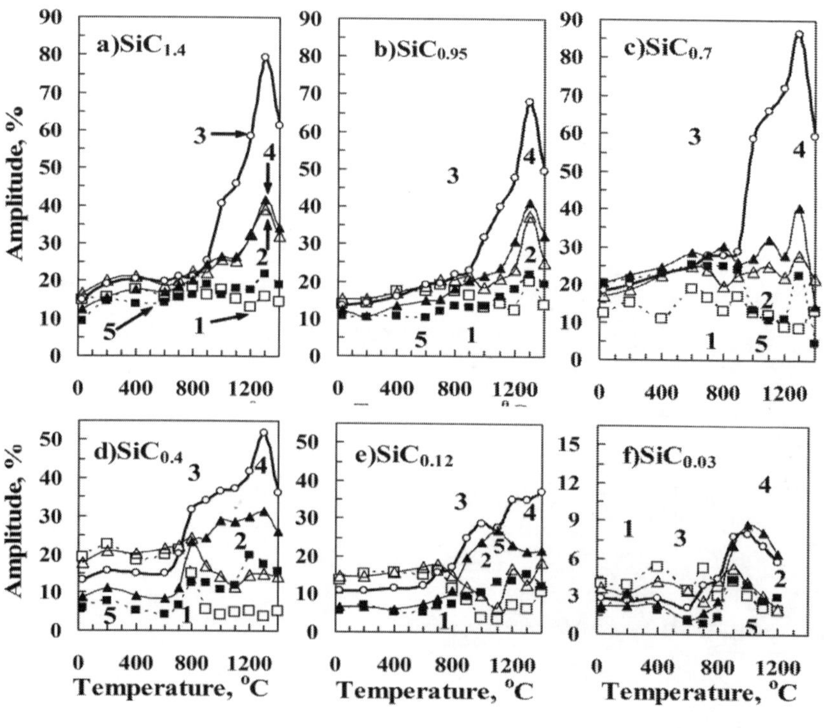

Figure 13. Effect of the annealing temperature on the IR transmission amplitude at wavenumbers of (1-□) 700 cm^{-1}, (2-Δ) 750 cm^{-1}, (3-○) 800 cm^{-1}, (4-▲) 850 cm^{-1}, and (5-■) 900 cm^{-1} under normal incidence of IR radiation on the sample surface: a) $SiC_{1.4}$; b) $SiC_{0.95}$; c) $SiC_{0.7}$, d) $SiC_{0.4}$, e) $SiC_{0.12}$; f) $SiC_{0.03}$.

The temperature range 800-1000°C

Accordingly to Figs. 6-8 and 12 almost whole ion implanted layer takes place in the process of crystallization of Si and SiC in this temperature range. The probability of over-barrier mechanism of the formation of Si- and SiC crystallites, seemingly, is increased as is seen from the significant increase of the IR transmission amplitude at 800 cm^{-1} (Fig.7, curves 1, 1'). The flat net of clusters and the chains of them in a great extent are disintegrated. The intensive decay of the infrared active non-tetrahedral Si–C-bonds is taken place and, the sesquialteral- and, partially, the double silicon-silicon bonds simultaneously with the infrared inactive single bonds can disintegrated as well. Seemingly, the energy of the phonons may be sufficient for the disintegration of the infrared inactive C–Si-, C–C- and even C=Si-bonds. Simultaneously, the formation of the most energy favorable tetrahedrally oriented bonds of Si- and SiC-crystallites is taken place (Figs. 6, 7, curves 1, 1'; Fig.12, curve 3). The continious process of the formation of the infrared active clusters absorbing on the frequencies close to 700–800 cm^{-1} is taken place in the layer and, the concentration of these clusters with the increasing of the annealing temperature is increased. The absorption at 800 cm^{-1} begins significantly predominate over the ones at another frequencies. That leads to the significant increase of the IR transmission amplitude at this frequency in comparison with the increase of amplitude at another ones (Fig.12, curve 3) and, that is perceived as a frequency shift of the IR absorption maximum (Fig.4). The increase of concentration of clusters absorbing on frequencies higher than 800 cm^{-1} is observed, too (Fig. 12, curves 4, 5).

So, the increase of the area under the IR transmission curve in the temperature range 800-1000°C is caused, mainly, by the absorption at frequency of 800 cm^{-1}, i.e. by bonds characteristic to the SiC-crystallites and, by bonds absorbing on frequencies both more and less than 800 cm^{-1} as well. In the case of the (100) oriented substrate the number of tetrahedral Si–C-bonds reached at 1000°C some maximum and does not change up to 1200°C, whereas in the case of orientation (111) it increases going smoothly in the range 900-1300°C. Comparison with the data in Fig.13 shows that such flat areas of curves at these temperatures are typical for the layers $SiC_{0.7}$, and especially for $SiC_{0.4}$. The total dose of implanted ions of carbon in the case of $SiC_{0.7}$ was $D(SiC_{0.7}) = 4.54\times10^{17}$ cm^{-2}, and $D(SiC_{0.4}) = 2.72\times10^{17}$ cm^{-2}, and is comparable to the dose of carbon ions with an energy of 40 keV for the considered Gaussian distribution of carbon: D (40 keV) = 3.56×10^{17} cm^{-2}. The halfwidth of IR-spectrum maximum (Fig.8) in the range 800-1000°C is rapidly decreased. That is an evidence of a significant ordering of the ion implanted layer structure caused by the formation of Si- and SiC-crystallites. It goes more intensively in case of (100) orientation of substrate.

The temperature range 1000-1100°C

In spite of the fact that the formation of new SiC crystallites in the implanted layer at these temperatures is not taken place (Fig.7, curve 1, 1'; Fig.12 and 16, curves 3), the decrease of area of SiC-peak of IR transmission curve is significant (Fig.6, curves 1, 1'). As it was shown earlier (Fig.11), the dimensions of Si- and SiC-crystallites are enlarged with the increase of the annealing temperature. Thereby, we believe that in this temperature range the uniting of small crystallites of Si and SiC in the larger ones is taken place, resulting the frequency shift of LO-phonon peak in IR spectrum to a higher frequency (Fig. 4), as well as the growth of its amplitude (Fig. 7, curve 3). When combining the crystallites, in the area of their union the decay of the both optically active and inactive clusters is taken place. This explains the decrease in the amplitude of the infrared transmittance for clusters absorbing at frequencies of 800-900 cm^{-1} (Fig. 12) and the corresponding decrease in the area (Fig. 6). Apparently, this temperature is insufficient for the decay of clusters C=Si and C=C and, as a result the formation of new SiC crystallites is not observed. However, a volume of polycrystalline Si is continuously increased due to the disintegration of Si-Si, Si=Si bonds in regions with low concentration of carbon (Fig.1, regions I and II).

In the case of the substrate orientation Si(111), the SiC-peak area is reduced to a lesser extent, and is larger after annealing at 1100°C than the peak area for the substrate Si(100) due to the fact that the number of tetrahedral bonds and crystallites continues to grow.

The temperature range 1100-1200°C

Both the area of Si–C-peak (Fig. 6) and its half-width (Fig. 8) increase in this temperature interval, while the volume of the polycrystalline SiC is unchanged (Fig. 12, curve 3 and Fig. 7, curve 1), as the further growth of SiC crystallite size due to their association (Fig. 4, curves of LO-phonons) is taken place. The growth of the area and half-width of Si–C-peak are caused by the formation of new optically active clusters absorbing at frequencies of 750-900 cm^{-1} (Fig. 12) due to the decay of stable clusters. As a result, the ion-implanted layer degrades the structure (increase in half-width of the peak).

The temperature range 1200-1250°C

No significant changes in the formation of new infrared active clusters over this range are taken place. The decaying clusters with short Si–C-bonds, absorbing at frequencies of 800–900 cm^{-1} (Fig. 12, curves 1-3) are converted into clusters with long bonds, absorbing at 700 and 800 cm^{-1} (Fig. 12, curves 4 and 5). Therefore, the area of Si–C-peak is not changed (Fig.6) for both orientations of substrate (100) and (111). The annealing temperature 1250°C may be sufficient for the decay of sesqui- and double Si–C-bonds. As a result, the concentration of tetrahedrally oriented Si–C-bonds increases, the atoms are combined into crystallites of SiC, and the amplitude of the IR spectrum at 800 cm^{-1} increases (Fig. 12, curve 3). There is a further streamlining of the structure of the ion-implanted layer (Fig. 8), both due to the formation of new crystallites of SiC, as well as due to an increase in their size (Fig. 4, curve of LO-phonons).

The temperature range 1250-1300°C

Seemengly, there may be competing processes here. Firstly, in the case of the substrate Si(100) in this interval there is a decay of a large number of tetrahedral bonds (Fig. 7, curve 1), which is the main reason for reducing the area of SiC-peak at 1300°C (Fig. 6, curve 1). At the same time the amplitude of the LO-phonon is decreased (Fig. 7, curve 3) and the half-width of the peak is increased (Fig. 8, curve 1), indicating a deterioration of the structure. In contrast, in the case of the substrate Si(111) the peak area (Fig. 6, curve 1') and the amplitude at 800 cm^{-1} (Fig. 7, curve 1') are increased, and this is accompanied by a decrease in the half-width of the TO-phonon peak and an increase in amplitude of the LO-phonon peak (Fig. 7, curve 3'), i.e. by the improving of the layer structure. We assume that there may be two dominant mechanism of the influence of substrate orientation on the layer structure at high temperatures. During the recrystallization of the damaged layer in the interface "the SiC film – Si substrate", both a destruction of the silicon crystallites and the uniting of their atoms with the substrate are taken place. The difference of the recrystallization of the substrate Si(100) may be the appearance of forces and conditions for the destruction of the defective crystallites of silicon carbide. This leads to a decrease in amplitude at 800 cm^{-1} and increase the half-width of the peak due to the appearance of non-tetrahedral Si–C-bonds. The second mechanism may be associated with different concentrations of carbon near the surface of silicon. After implantation into (100) oriented Si substrate, the carbon-riched surface layer is much thicker than in case of Si(111) substrate, so the sublimation and desorption of carbon at high temperatures will lead to a significant decrease in the amplitude values of the SiC-peak at all frequencies. Ie, the experiments to study an influence of substrate orientation on the desorption of implanted carbon are necessary.

The temperature range 1300-1350°C

Despite the increase in the desorption of carbon, the area of SiC-peak (Fig. 6), as well as the amplitude at all frequencies of spectra are increased in case of the film on the substrate Si(100) (Fig. 12, curves 1-5). The atoms of clusters, which formed due to the destruction of defective crystallites of SiC in the previous temperature range, re-unite again in form of the crystallites of SiC, as well as in form of optically active clusters, which absorb at frequencies near 800 cm^{-1}. In addition, stable clusters with Si=C, Si≡C and C=C bonds are disintegrated (Fig. 9). Growth of SiC crystallite size leads to a frequency shift of LO-phonons peak in the short-wavelength region (Fig. 4). Since in an isolated system unacceptable the processes occurring with increasing free energy, the uniting of two crystallites occurs, if is accompanied by a gain in energy in comparison with the energy expended in their decay. In the case of the orientation of the substrate Si(111) a decrease of the area of SiC-peak (Fig. 6, curve 1'), as well as the amplitude at 800 cm^{-1} (Fig. 7, curve 1') occur due to increased desorption of carbon.

The temperature range 1350-1400°C

Although at these temperatures the decay of optically inactive clusters, formation of both new crystallites and optically active SiC-clusters should be the greatest, nevertheless the growth of area and the amplitude of the SiC-peak of IR spectrum is not observed. On the contrary, they decrease (Fig. 6and 7), which can be explained to the dominant influence of sublimation and desorption of carbon. The volume of polycrystalline SiC is also reduced, which is accompanied by a decrease in the amplitude of LO-phonons (Fig. 7, curve 3). A further ordering of the structure of ion-implanted layer occurs, as evidenced by the decrease in the peak half-width of IR spectrum.

In conclusion, it is necessary to note that the quantity of absorbing Si–C-bonds in the silicon layer with Gaussian distribution of implanted carbon reaches a maximum at 1000°C for (100) oriented substrate and, at 1000 and 1250°C for (111) oriented substrate (Fig. 6). Most of the carbon atoms combine with atoms of silicon, forming the tetrahedrally oriented bonds of SiC (Fig. 12, curve 3). Seemingly, there are flat nets and chains of clusters (Fig. 9), which consist mainly of bonds Si–Si, Si = Si, C–Si, C–C and, this temperature is sufficient for their decay. Some significant part of the carbon atoms form bonds of higher order, which decay at temperatures of 1200-1400°C and above. At high temperatures, 1300-1400°C (Figs. 6 and 7) occur intense desorption processes of carbon.

A shape of IR transmission peak

All presented spectra (Fig.2) have shape different from simply dispersive spectrum (theoretically calculated). The transmission band both left and right from the transmission peak are perceptible asymmetric and, one can not describe it's shape by a simple analytical function. The shape asymmetry is decreased with the annealing temperature increasing and, it is minimal in the temperature range 1300–1350°C. The further increase of the annealing temperature leads to the increase of asymmetry, too. The contour of the transmission peak for TO-phonons at perpendicular incidence of electromagnetic radiation on sample surface after annealing at 1350°C is most close to the dispersive one.

Obviously, the asymmetry of IR transmission contour is related with the presence of the infrared active clusters in the ion implanted layer, and the concentration of clusters is minimal when the asymmetry is minimal, i.e. at 1350°C. In this relation, a largest area of SiC-peak corresponds to maximum amplitude of absorption at wavenumber 800 cm^{-1} (Figs.7 and 8). As is seen from amplitude values in Fig.12 (curves 1, 2, 4, 5), there is a certain quantity of the non-tetrahedral optical active clusters at 1350°C. Seemingly, a presence of very stable optical inactive clusters, which are not disintegrated even at the melting point of Si, is possible.

Measurement of a conduction type of carbon implanted silicon layer

The (100) oriented substrates of n- and p-Si of dimensions 7×5×0.3 mm^3 with resistivities 4–5 Ом см have been implanted by carbon ions with values of energy 40 keV and dose 3.56×10^{17} cm^{-2} to determine a type of conduction. After implantation, the samples have been isochronously annealed in vacuum over the temperature range from 200 up to 1200°C with step 200°C for 30 min. A surface layer of the annealed samples have been removed by etching in an acid mixture $HF:HNO_3$ in composition of 1:10. A type of conduction of the implanted surface has been determined using thermo-emf after each 0.5 mm along both horisontal and vertical directions. The thermo-emf have fixed with approximately equiprobability the both n- and p-type of conduction on n-Si substrates, while on the p-Si substrates the thermo-emf have shown the p-type of conduction only. We believe that p-type of conduction on the n-Si substrates is provided by the SiC crystallites. The conduction of the Si crystallites is similar to the conduction of the substrate. If the substrate is p-Si, so the both Si- and SiC-crystallites have the p-type of conduction. So, the synthesized SiC-crystallites have p-type of conduction independently from the type of substrate.

3.2. Investigation Of High-Temperature Instability Of Solid Sic Films Synthesized By Ion Implantation

As stated in paragraph 2, for construction of a rectangular profile of the distribution of carbon atoms in the silicon, the implantation of carbon ions of different energies and doses in the second group of single-crystal silicon wafers of n- and p-type conductivity was carried out sequentially in the order according to Table 2. The doses of ions were chosen in such a way to obtain a layer $SiC_{0.7}$ with the ratio of the concentrations of carbon and silicon atoms through a depth of about $N_C/N_{Si} = 0.7$. Postimplantation annealing of the samples was performed in a vacuum in the temperature range 200-1200°C for 30 min with a step 200°C. In some cases, to compare also were analyzed the samples with films $SiC_{0.95}$.

3.2.1 Influences Of Annealing, Sputtering And The Film Composition Changes During High Dose Implantation On The Thickness And Shape Of The Distribution Profile Of Carbon Atoms

Fig. 14 shows the calculated profile N_C(Gibbons) of distribution of carbon atoms through the depth of silicon for the energies and doses of ions according to Table 2, which is the sum of Gaussian distributions constructed with the use of $R_p(E)$ and $\Delta R_p(E)$ by [21] (LSS) in accordance with the expression:

$$N(x) = \frac{D}{\Delta R_p (2\pi)^{1/2}} \exp\left[-\frac{(x - R_p)^2}{2\Delta R_p^2}\right] \qquad (5)$$

where x – the distance from the surface.

Table 2. Values of energy, E, dose, D, projected range, $R_p(E)$, and straggling, $\Delta R_p(E)$, for $^{12}C^+$ ions in Si, used for constructing a rectangular distribution profiles $SiC_{0.7}$ and $SiC_{0.95}$.

E, keV		40	20	10	5	3
$D(SiC_{0.7})$, 10^{17} cm^{-2}		2.80	0.96	0.495	0.165	0.115
$D(SiC_{0.95})$, 10^{17} cm^{-2}		4.48	1.54	0.792	0.264	0.184
N_C(Gibbons) profile [21]	$R_p(E)$, nm	93.0	47.0	24.0	12.3	7.5
	$\Delta R_p(E)$, nm	34.0	21.0	13.0	7.0	4.3

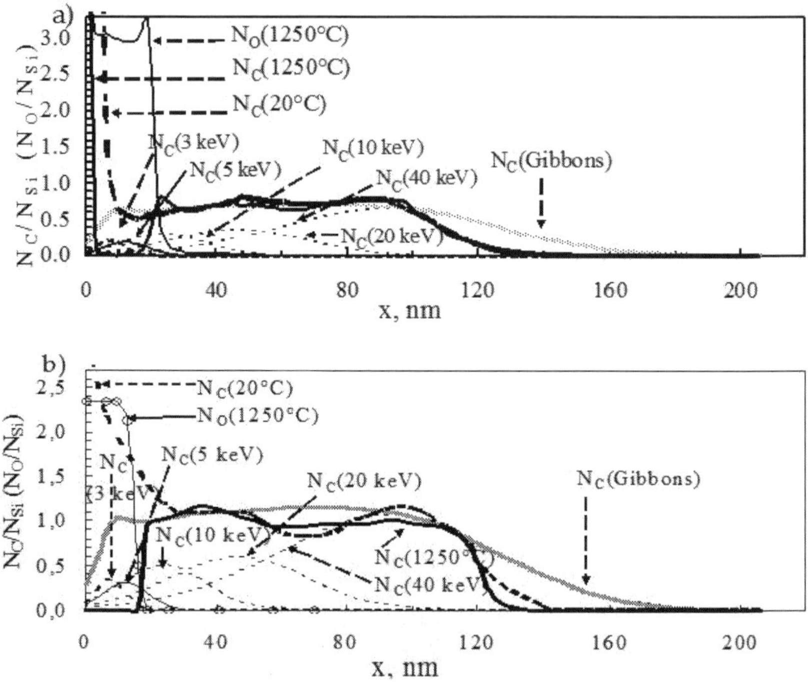

Figure 14. distribution profiles in Si produced by ion implantation (see Table 2). (a) $SiC_{0.7}$; (b) $SiC_{0.95}$;N_C (Gibbons) is the profiles calculated according to [21], where N_C(Gibbons) = N_C(40 keV) + N_C(20 keV) + N_C(10 keV) + N_C(5 keV) + N_C(3 keV). N_C(20°C), N_C(1250°C) and N_O(1250°C) are the Auger profiles of carbon and oxygen, respectively, in a layer after high-dose implantation and annealing at 1250°C for 30 min.

Fig. 14 also shows the experimental curves (Fig. 14, curves N_C(20°C), N_C(1250°C) and N_O(1250°C)), obtained by Auger electron spectroscopy, showing the ratio of the concentrations of carbon and oxygen atoms to silicon (N_C/N_{Si} и N_O/N_{Si}) through the depth of the sample after implantation (20°C) and annealing at 1250°C for 30 min in an argon atmosphere containing some oxygen. These distributions are constructed through a depth, taking into account the condition that the number of carbon atoms in silicon and, consequently, the integrals and the area under the curves N_C(Gibbons) and N_C(20°C) must be equal one another at a first approximation. The areas under the curves were equal to $S_G = S_{20°C} = \int(NC/NSi)dx\int(NC/NSi)dx$ = 90 units (or 100%), and after annealing at 1250°C: $S_{1250°C}$ = 71 units (or 79.4%) due to the formation of silicon oxide layer. When evaluating the number of carbon atoms in a thin surface region (8 nm), where N_C/N_{Si} is very great due to the low content of silicon atoms ($N_{Si} << 5\times10^{22}$ cm^{-3}), an approximation was made that the N_C/N_{Si} does not exceed 2.3 ($N_{graphite} = 11.6\times10^{22}$ cm^{-3} and $N_{silicon} = 5\times10^{22}$ cm^{-3}). At the same time, the area under the profile curve for the region x> 22.2 nm were estimated S_G = 78 units, $S_{20°C}$ = 66 units and $S_{1250°C}$ = 65 units. That is, the areas under the profile curves before and after annealing for x > 22.2 nm are almost equal, but less than calculated value, since part of the carbon atoms after implantation was concentrated near the surface (x <8 nm), and during annealing occur desorption of carbon from the layer (x<22.2 nm) and the formation of silicon oxide. The interface "the SiC film - Si substrate" in the experiment was more abrupt than it was expected. After annealing for 30 minutes, almost 20% of the total number of carbon atoms desorbed from the carbon-rich surface layer of the film. Fig. 14 shows that the average concentration of carbon and oxygen were: N_C/N_{Si} = 0.7 in the depth 22.2 < x < 110 nm and the $N_O/N_{Si} \approx 3.0$ at the surface layer x < 22.2 nm. In this case there is penetration of oxygen atoms into the layer up to 30 nm.

Some difference between the shape of the experimental and calculated curves of the profile is observed (Fig. 14). The distribution N_C(Gibbons) was made without taking into account the effects of sputtering and composition changes in the layer by high dose implantation. Accounting for the effect of surface sputtering during high dose of implantation of carbon ions (E = 40 keV, D = 2.8×10^{17} cm^{-2}) allows to assume the displacement of profile further into the layer with increasing dose, to some expansion of the profile and, consequently, to reduce the carbon concentration at the peak of the distribution in comparison with the calculated value. However, changing the composition of the single-crystal silicon substrate up to a mixture of C and Si atoms during the implantation suggests the formation of a significant amount of double and tripple Si–C- and C–C-bonds, which are more strong than the Si–Si-bonds, as well as the formation of stable carbon and carbon-silicon clusters. This results a decrease of $R_p(E)$ and $\Delta R_p(E)$ during implantation.

The decrease of $R_p(E)$ decreases the influence of surface sputtering on the position of the distribution maximum of carbon atoms, i.e., the maximum should remain nearly on the same depth. Moreover, the decrease of $\Delta R_p(E)$ will increase the carbon concentration at the maximum of peak and a more sharp decrease in concentration in the direction to the substrate and to the surface. This will cause a decrease in the depth of the interface "film SiC – the substrate Si», which becomes more sharp with increasing dose, as well as an occurrence of depression between peaks 40 and 20 keV, and possibly between 20 and 10 keV. The appearance of depression between peaks 40 and 20 keV in Fig. 14 for layers $SiC_{0.7}$ and $SiC_{0.95}$ may be due to these reasons.

The surface sputtering during the implantation of carbon ions with energies of 10, 5 and 3 keV is more intense with decreasing ion energy. This should lead to an increase in carbon concentration near the surface due to shift of the distribution maxima of these ions one to another (N_C(3 keV) and N_C(5 keV) in the direction of N_C(10 keV)). As a result, significant increase in the concentration of carbon in the surface layer is observed.

It is seen in Fig.14b that the average concentration of carbon and oxygen were: N_C/N_{Si} = 0.95 in the depth of the layer from 20 to 110 nm and the $N_O/N_{Si} \approx 2.33$ at the surface layer up to a depth of about 20 nm. At the same time observed the penetration of oxygen up to 80 nm to the depth. It was found for a layer $SiC_{0.95}$ that $S_G = S_{20°C} = \int (N_C/N_{Si})dx = 144$ units (or 100%), and after annealing at 1250°C: $S_{1250°C}$ = 103 units (or 71.3%). Thus, it appears that after annealing for 30 minutes, almost 30% of the carbon atoms desorbed from the surface layer of the film. At the same time the appearance of a layer of oxygen atoms at the surface is revealed. The area under the profile curve for the region x > 25 nm were estimated S_G = 121 units, $S_{20°C}$ = 99 units and $S_{1250°C}$ = 95 units. i.e., the area under the curves of the profile before and after annealing for region x > 25 nm are similar in magnitude, but again less than calculated one, since part of the carbon atoms after implantation was concentrated near the surface (x < 19 nm), and after annealing, there was desorption of carbon from layer (x <25 nm) and the formation of silicon oxide. The interface "the SiC film – Si substrate" in the experiment also was sharper than the expected one.

The presence of a sharp interface "the SiC film – Si substrate" permits to suppose that is possible to obtain promising results on the measurement of film thickness by X-ray reflectometry, although this method is typically used for films deposited with a very sharp interface "a film - substrate" and for ion-implanted layers usually does not apply. The parameters of the $SiC_{0.7}$ film by this method were investigated at small grazing angles θ by recording the angular dependence of the reflection coefficient using two spectral lines CuK_α (0.154 nm) and CuK_β (0.139 nm) on the installation "ComplexXRay C6" [61]. The oscillations of intensity were observed, assigned to the interference of X-ray reflections in the layers $SiC_{0.7}$ and SiO_2 (Fig. 15).

The first maximum of reflection with intensity I_1 = 93207 pulses at an angle of 2θ = 0.418° is observed (Fig. 15). The angle of total external reflection is evaluated as an angle where the intensity is equal to a half of the first maximum ($I = I_1/2$ = 46603 pulses), ie $2\theta_c$ = 0.449°, or θ_c = 0.2245° = 3.918 mrad. Using the Henke program is determined that this value of θ_c corresponds to the value of film density 2.37 g/cm^3, which is close to the density of cristobalite (SiO_2) 2.32 g/cm^3. Further, with increasing of the incidence angle, the intensity of reflection increases again up to I_2 = 76831 pulses and that indicates the presence of a more dense structure. If the intensity falls up to the value $I = I_2/2$ = 38415 pulses, the value $2\theta_c$ = 0.486°, the critical angle is equal to θ_c = 0.243° = 4.241 mrad, which corresponds to a density 2.77 g/cm^3 and is close to the density of quartz (SiO_2) 2.65 g/cm^3. As shown in Fig. 15, then there is a second increase in intensity up to I_3 = 34416 pulse which corresponds to a denser structure. If the intensity falls up to the value $I = I_3/2$ = 17208 pulse, the value $2\theta_c$ = 0.526°, and θ_c = 0.263° = 4.590 mrad. This corresponds to a density ρ = 3.25 g/cm^3, which is close to the density of silicon carbide - 3.2 g/cm^3.

The layer thickness is determined by the formula $2d \sin \theta = \lambda$, or taking into account the small values of θ: $d = \lambda/2\ \theta$ nm, where λ - the wavelength of CuK_α (0.154 nm) or CuK_β (0.139 nm) radiation, and $2\theta_{av}$ was determined as an average from several (j - i) peaks (Table 3). To determine the thickness, four narrow peak of SiC, and two broad bands of SiO_2, probably from two different phases - cristobalite and quartz (Fig. 15), were used. The second broad band consists of 3 bands. The thickness of the resulting system ($SiO_2 - SiC_{0.7} - Si$) was about 100 nm.

Similar measurements for the $SiC_{0.95}$ layer also led to the observation of the intensity oscillations of X-ray reflections. The first maximum of reflection with intensity I_1 = 98703 pulses at an angle of 2θ = 0.396° is observed (Fig. 16). The critical angle of total external reflection is evaluated as an angle where $I = I_1/2$ = 49352 pulses, but in this case in the position of a minimum I = 53961 pulse, i.e. $2\theta_c$ = 0.458°, and θ_c = 0.229° = 3.979 mrad. This angle corresponds to the film density 2.46 g/cm^3, which is close to the density of optical glass 2.51 g/cm^3. Further, with increasing of incidence angle, the intensity of reflection is again increased up to I_2 = 57255 pulses, which indicates the occurrence of a more dense structure. If the intensity falls up to the value $I = I_2/2$ = 28627 pulses, the value $2\theta_c$ = 0.510°, the critical angle is equal to θ_c = 0.255° = 4.451 mrad. This corresponds to a density 3.06 g/cm^3, which is close to the density of silicon carbide - 3.2 g/cm^3.

To determine the thickness of the layers, four narrow peaks of SiC, and two broad bands of SiO_2 (Fig. 16) were used.

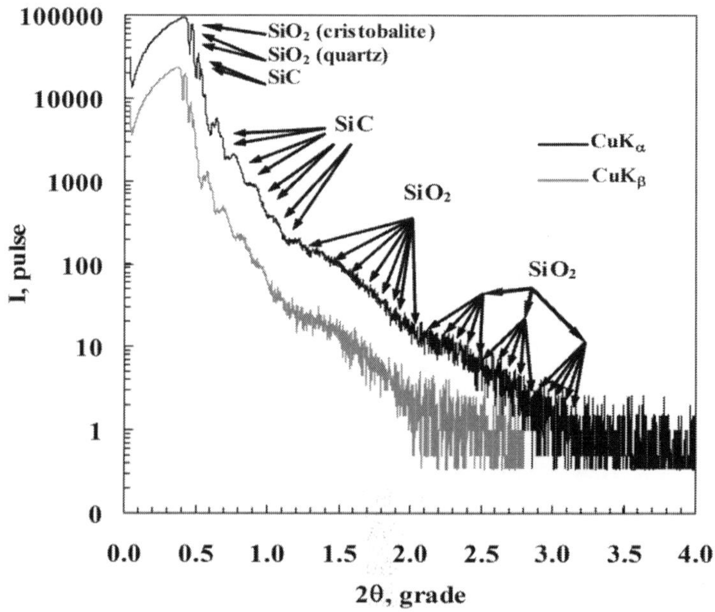

Figure 15. X-ray reflectometry using two spectral lines CuK_{α} (0.154 nm) and CuK_{β} (0.139 nm) ("ComplexXRay C6") of parameters of the $SiC_{0.7}$ films, synthesized by multiple implantation of carbon ions with energies of 40, 20, 10, 5 and 3 keV into silicon, after annealing at 1250°C.

The thickness of the system ($SiO_{2.33}$ – $SiC_{0.95}$ – Si) was 117 nm, which was comparable to the estimated thickness of the film. SiO_2 peaks are visible not clearly. Perhaps this is because the annealing temperature was taken at 100°C lower (1150°C).

Table 3. Determination of the thickness of the layers in the system (SiO_3 – $SiC_{0.7}$ – Si) by X-ray reflectometry according equation 2 d· sin θ = λ.

Layer	$(2\theta)_j$	$(2\theta)_i$	j – i	$2\theta_{av} = [(2\theta)_j - (2\theta)_i]/(j-i)$	λ	d = λ/2θ, nm
SiC	1.138	0.598	4	0.135	0.15405	65.4
SiO_2	3.154	2.012	3	0.38	0.15405	23.2
SiO_2	2.012	1.178	1	0.83	0.15405	10.6

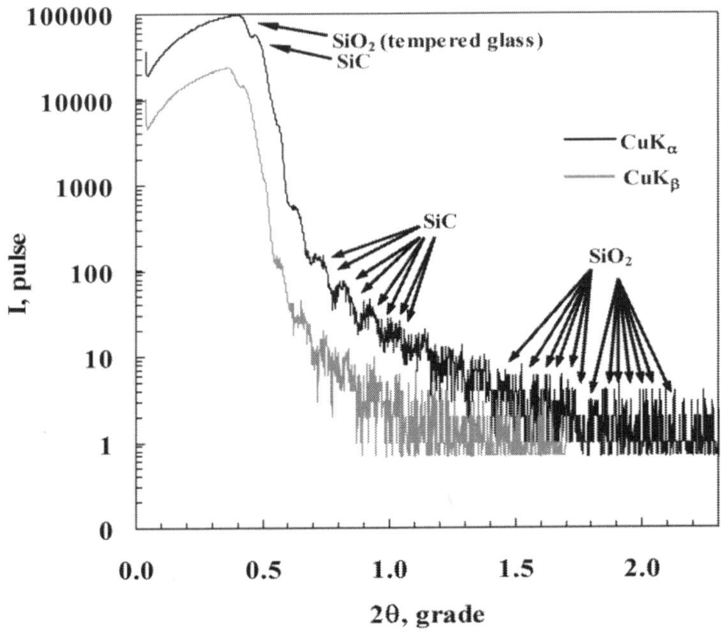

Figure 16. X-ray reflectometry of parameters of the films $SiC_{0.95}$, synthesized by multiple implantation of carbon ions with energies of 40, 20, 10, 5 and 3 keV into silicon after annealing at 1150°C.

3.2.2 The Study Of The High-Temperature Instability Of Solid $Sic_{0.7}$ Films Synthesized By Ion Implantation

In Fig. 17, the IR transmission spectra of the homogeneous $SiC_{0.7}$ films, synthesized on substrates of Si(100) with resistivity 4–5 Ohm cm (a) and Si(111) with resistivity 10 Ohm cm (b), subjected to isothermal annealing at the temperature 1200°C for several hours in an atmosphere of inert gas (Ar), are presented. Comparing these two figures (a) and (b), one can see that the nature of the SiC films formed on substrates with different crystallographic orientations, are different. This difference manifests itself in the amplitudes and half-widths of the peaks corresponding to the excitation of both transverse and longitudinal optical lattice oscillations of SiC (TO- and LO-phonons).

Table 4. Determination of the thickness of the layers in the system ($SiO_{2.33}$ – $SiC_{0.95}$ – Si) by X-ray reflectometry.

Layer	$(2\theta)_j$	$(2\theta)_i$	j – i	$2\theta_{av} = [(2\theta)_j - (2\theta)_i]/(j-i)$	λ	$d = \lambda/2\theta$, nm
SiC	1.066	0.690	4	0.094	0.15405	93.9
SiO_2	2.19	1.426	2	0.382	0.15405	23.1

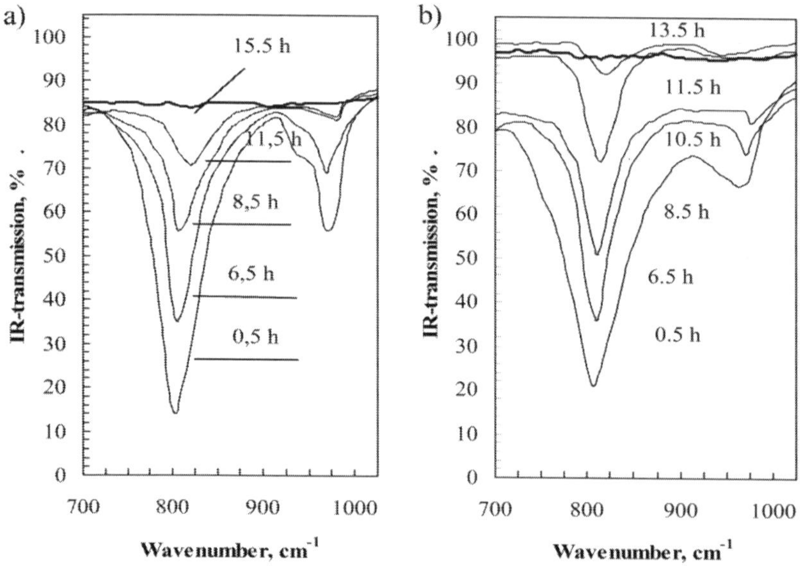

Figure 17. The dependence of the IR transmission spectra of implanted by $^{+}C^{12}$ ions Si on the annealing time at the temperature 1200°C: a) n-Si, the orientation of substrate Si(100), b) p-Si, the orientation of substrate Si(111).

After annealing at 1200°C for 30 min in the IR spectra an intense peak at 800 cm^{-1} which associated with the TO phonons of SiC, as well as a peak at 960 cm^{-1} corresponding to the LO-phonons of SiC, are observed. It is seen that in contrast to the spectra of film on (100) oriented Si substrate, the transmission spectra of oscillation modes of SiC of film on the (111) oriented silicon substrate are more blurred and the level of the transmission spectra of the two modes are superimposed on each other, and does not achieve the initial zero level in the wave number, equal to 915 cm^{-1}. This is caused by the half-width of these peaks (Fig. 17 and 18).

The narrowing of the peak (Fig. 18) up to 40 cm $^{-1}$ occurs as a result of intensive formation of the tetrahedrally oriented Si–C-bonds absorbing at 800 cm $^{-1}$, as well as the decay of bonds which absorb at frequencies different from the value of 800 cm^{-1}. Since the tetrahedral bond corresponds to the crystalline phase of silicon carbide, the narrowing of the peak of the IR spectrum with a minimum at 800 cm^{-1} is associated with the process of ordering of the implanted layer. As is seen from Fig.

18, for the $SiC_{0.7}$ layer the narrowing of the peak is more intense with increasing time of isothermal annealing up to 6.5 hours in the case of the (111) oriented substrate in comparison with (100) orientation. After annealing for 8.5 hours or more a further narrowing of the peak is slowing, indicating a complete processes of SiC lattice ordering. Thus, it was established that the annealing duration of the less than 6.5 hours at 1200°C is insufficient to form the structure of silicon carbide.

As is seen from Figs. 17 and 19, the amplitude values of the peaks with increasing of annealing time at 1200°C are reduced. This indicates a decrease in the total volume of silicon carbide due to disintegration of SiC and desorption of carbon. Since the amplitude of the SiC-peak of infrared transmission is proportional to the concentration of Si–C-bonds, the measurements of its value were made in the spectra after isothermal annealing at the temperature of 1200°C (Fig. 19). For (100) oriented silicon substrate of n-type conductivity, the amplitude of the TO- and LO-phonon peaks of the infrared transmission (Fig. 19, curves 2 and 4) after annealing for 0.5–6.5 hours were higher than the same for (111) oriented silicon, and then the decay of SiC in this layer becomes more intense. However, as is seen from Figs. 17 and 19, after annealing for 11.5 and 13.5 hours the disintegration of silicon carbide is almost finished for the $SiC_{0.7}$ layer on the (111) substrate, while for the (100) orientation is observed after annealing during up to 15.5 hours.

Since the amplitude of the infrared transmission at the wavenumber 800 cm^{-1} is proportional to the concentration of the tetrahedrally oriented Si–C-bonds, its magnitudes were measured in the spectra after isothermal annealing at the temperature of 1200°C (Fig. 20, curve 3). Assuming that the amplitude at any frequency is proportional to the number of the Si–C-bonds which absorb at this frequency, the amplitudes for the TO-phonons with wavenumbers 700, 750, 850 and 900 cm^{-1} (Fig. 20)in the case of incidence of IR radiation to the sample surface at an angle of 73° to the normal were also measured.

It should be also noted that the signal from the LO-phonons in the spectra of both types of substrate disappears earlier (Fig. 19) than the signal from the TO-phonons, and in particular, at (111) orientation of substrate. Thus, a gradual decrease in the amplitudes of the TO-and LO-phonon peaks of SiC in the IR transmission spectra of ion-synthesized SiC films at the increasing time of high-temperature annealing indicates the decay of the SiC structure, i.e. the instability of these films to such regime of treatment.

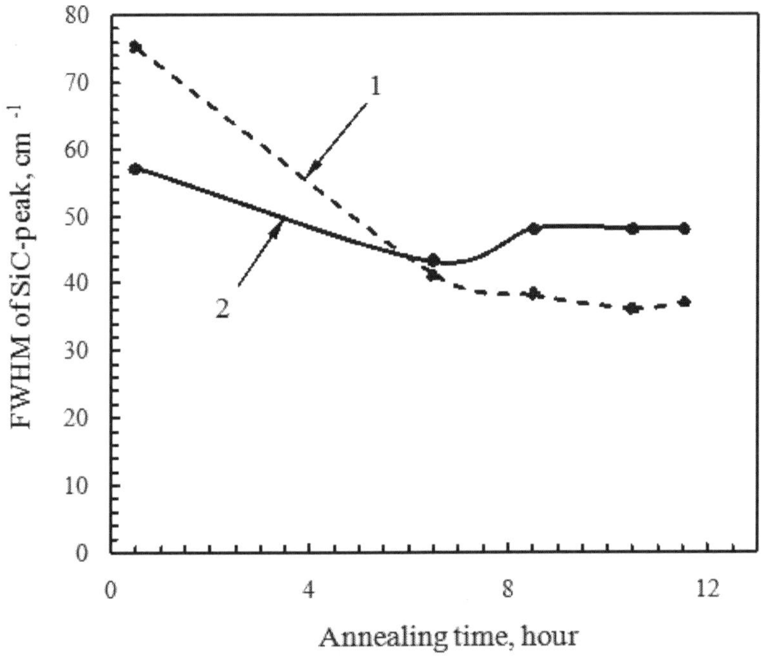

Figure 18. Dependence of the half-width of the TO-phonons peak of SiC of IR spectra on the annealing time at the temperature of 1200°C for $SiC_{0.7}$ layers: 1 - Si(111) substrate, 2 - Si(100) substrate.

It is seen in Fig. 20a, b (curves 3) that after annealing at 1200°C for 0.5 hour of the $SiC_{0.7}$ film on the Si (100) substrate, the amplitude at wavenumber 800 cm^{-1} is higher than the same for the Si(111) substrate (70 and 58%), indicating a higher content of the tetrahedrally oriented SiC-bonds. It is also seen that the number of SiC-bonds which are close to tetrahedral orientation and absorb at 750 and 850 cm^{-1}, in the case of the substrate Si (100) is lower after annealing for 0.5 h (Fig. 20, curves 2 and 4) in comparison with the substrate Si(111) due to their more intense transformation into the tetrahedral SiC-bonds. The nonlinear nature of the curve 3 in the region 0.5 – 8.5 hours may be due to the formation of tetrahedrally oriented SiC-bonds in the layer simultaneously with their decay at the surface. The saturation of this process after annealing during 6.5 hours results in a faster decrease in the number of bonds during further annealing.

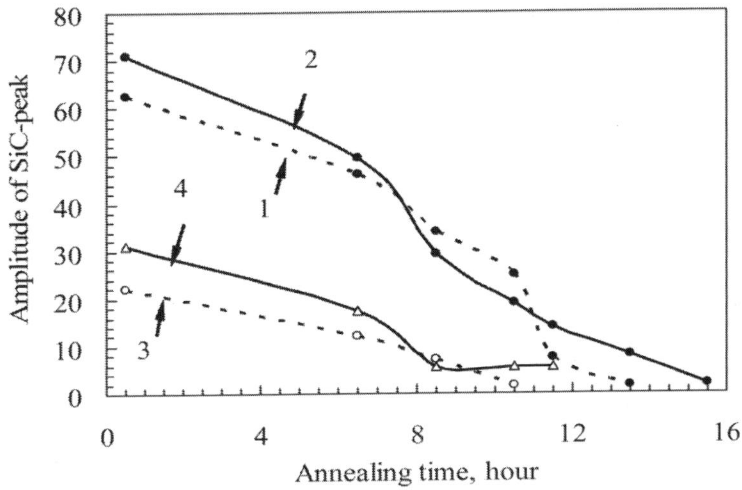

Figure 19. Amplitude of TO- and LO-phonon peaks of SiC of IR transmission versus the annealing time at the temperature of 1200°C for $SiC_{0.7}$ layers on silicon substrates of (100) and (111) orientation: 1 – Si(111), TO-phonon peak; 2 - Si(100), TO-phonon peak; 3 - Si(111), LO-phonon peak; 4 - Si(100), LO-phonon peak.

This process is clearly demonstrated on the time dependence of the area of SiC-peak, which is proportional to the total number of optically active Si–C-bonds (Fig. 21). Although the peak amplitude at the minimum of IR transmission (Fig.19, curve 2) and at 800 cm^{-1} (Fig.20, curve 3) for the Si substrates with (100) orientation are higher than in the case of (111) orientation, the value of area of SiC-peak for (111) was higher after annealing for 0.5 hours. This is due to a greater half-width of the peak caused by a significant amount of optically active Si–C-bonds close to tetrahedrally oriented, which absorb at 750 and 850 cm^{-1} and, probably due to the smaller amount of stable carbon silicon clusters in the film on (111) oriented silicon substrate. It is not contradict to the data for the SiC layer with Gaussian distribution of carbon in silicon, for which was shown that immediately after the implantation of carbon into the (100) and (111) oriented silicon at least 65% and 60% of carbon atoms are concentrated in optically inactive clusters, respectively (see 3.1.2, Fig. 6 "The temperature range 20-600°C"). The strong carbon clusters prevent the crystallization of SiC, and they are less susceptible to oxidation and prevent the penetration of oxygen into the SiC-layer. The SiC film on

(100) substrate has more quantity of stable clusters after implantation and, as a result the smaller value of area of SiC-peak and smaller amount of optically active Si–C-bonds after annealing for 0.5 hours and, it less susceptible to oxidation at 1200°C – 15.5 h (instead of 13.5 h for (111) Si substrate). The relatively rapid decay of close to tetrahedral Si-C bonds (Fig. 20b, curves 2 and 4) led to the decrease with higher speed of the number of optically active Si–C-bonds in the case of (111) orientation of Si substrate. In general the dependence of the reduction of optically active Si–C-bonds on the annealing time is linear. This implies that the rapid decay of close to tetrahedral Si–C-bonds was occurring, mainly, due to their transformation into a tetrahedral bond. The linear dependence indicates the homogeneity of the layer, and a rectangular profile of the distribution of carbon atoms in silicon, as well as the fact that the decay rate of silicon carbide does not depend on the depth of the oxidation front. Some decrease in the slope of angle of the curves at the end of the annealing is occurring due to oxidation of the interface "the SiC film – Si substrate".

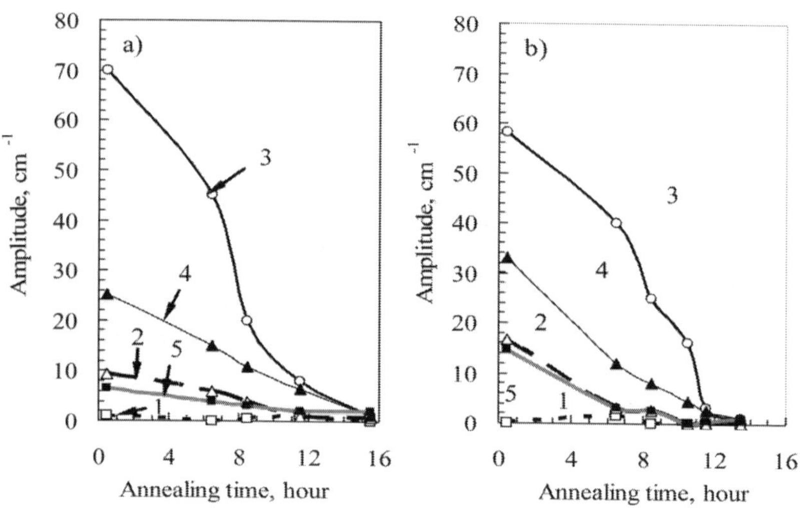

Figure 20. The amplitude of the infrared transmittance at fixed wavenumbers versus the duration of isothermal annealing of the $SiC_{0.7}$ layer (angle of incidence of infrared rays on the sample is 73° from the normal): 1 – 700 cm^{-1}, 2 – 750 cm^{-1}, 3 – 800 cm^{-1}, 4 – 850 cm^{-1}, 5 – 900 cm^{-1}; a) the substrate orientation Si (100), b) the substrate orientation Si (111).

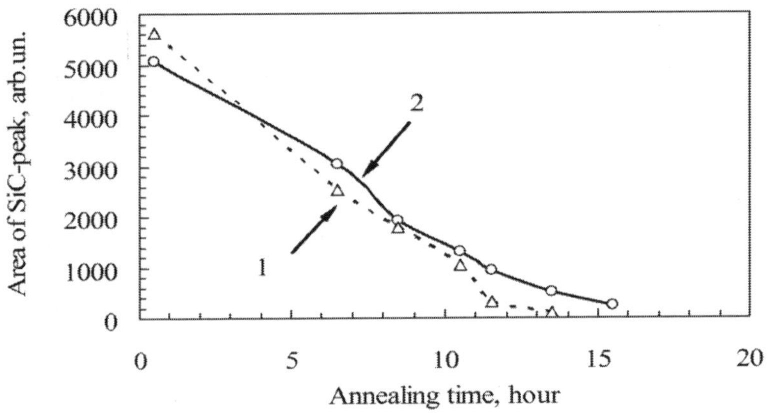

Figure 21. The area of the TO-phonon peak of SiC in spectra of the IR transmission versus the annealing time at the temperature of 1200°C for the $SiC_{0.7}$ layers (angle of incidence of infrared rays on the sample surface - 73° from the normal): 1 - the orientation of the substrate Si(111), 2 – the orientation of the substrate Si(100).

The value of the wave number of the minimum of SiC-peak of the infrared transmission (the position of the minimum of peak) defines a prevailing kind of optically active bonds which absorb at this wavenumber at this temperature. For the considered layers after annealing at 1200°C for 0.5 h, the transmission peaks with values of minimum at wavenumbers 803 and 806 cm^{-1}, characteristic of crystalline silicon carbide, are observed (Fig. 22). It was found that with increasing annealing time, the position of the peak minimum varies smoothly and moved in the direction of the wave number increase (Fig. 17 and 22).

In our opinion, the frequency shifts of SiC-peak upward indicate a decrease in crystallite size of SiC. We have previously identified size effects, which manifested in the influence of the crystallite sizes of silicon carbide on its optical properties. It was shown (Fig.23) that the differences of the $SiC_{0.03}$, $SiC_{0.12}$ and $SiC_{0.4}$ layers with low carbon concentration from the $SiC_{1.4}$, $SiC_{0.95}$ and $SiC_{0.7}$ layers with high carbon concentration are manifested in the absence of LO-phonon peak of SiC in the IR transmission spectra and in a shift at 1000°C of minimum SiC-peak for TO-phonons in the region of wave numbers higher than 800 cm^{-1}, characteristic for the tetrahedral bonds of crystalline SiC, which is caused by small sizes of SiC crystallites (≤ 3 nm) and by an increase of contribution in the IR absorption of their surfaces, and the surfaces of Si crystallites containing strong short Si–C-bonds as well.

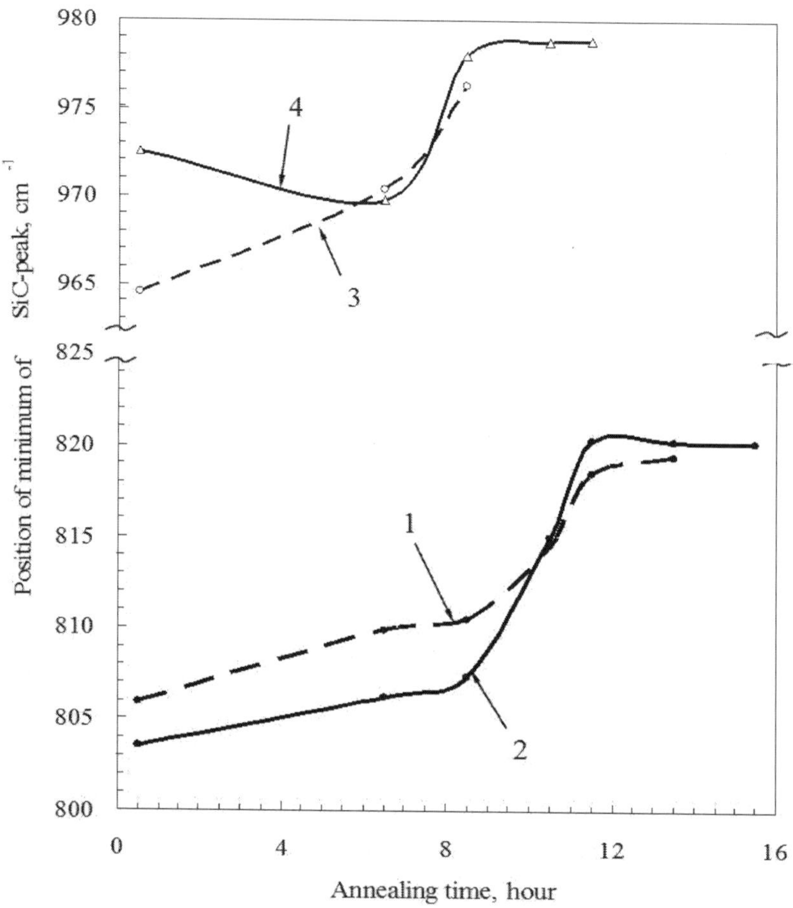

Figure 22. Wave number of the minimum of SiC-peak of IR transmission versus the annealing time at 1200°C for the $SiC_{0.7}$ layers on silicon substrates of (100) and (111) orientations: 1 - Si (111), TO-phonons, 2 - Si(100), TO-phonons, 3 - Si(111), LO-phonons, 4 - Si(100), the LO-phonons.

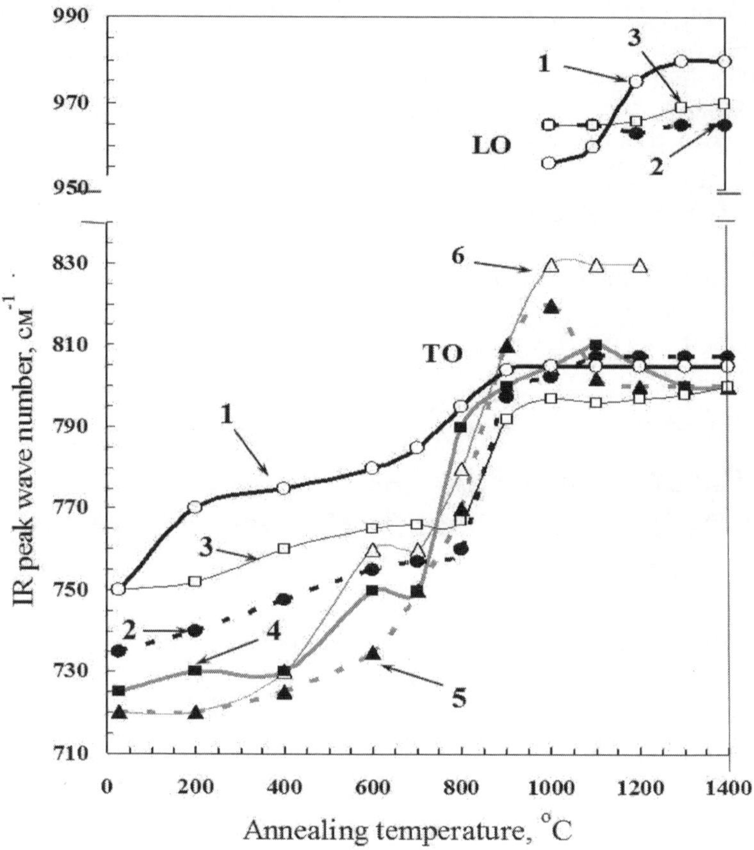

Figure 23. Wavenumber of the IR transmision peak for TO- and LO-phonons SiC as a function of the annealing temperature: 1 - $SiC_{1.4}$, 2 - $SiC_{0.95}$, 3 - $SiC_{0.7}$, 4 - $SiC_{0.4}$, 5 - $SiC_{0.12}$, 6 - $SiC_{0.03}$.

In this case (Fig. 22), the increase of annealing duration of $SiC_{0,7}$ layer leads to both the shift of the minimum of the IR transmission peak up to 820 cm^{-1}, and the reduction of the amplitude of the LO-phonon peaks and their subsequent disappearance, although $SiC_{0.7}$ is considered as the layer with a high concentration of carbon. At the same time more intense process of the shift of the minimum of SiC-peak occurs after annealing for longer than 8.5 hours, which leads to the disappearance of the peak of LO-phonons. This can occur when the penetration of oxygen deep into the layers, their interaction with the carbon atoms on the surface of the crystallites of silicon carbide with the formation of molecules of CO/CO_2. Desorption of carbon atoms causes a decrease in the size of the SiC crystallites and their disintegration. With increasing duration of annealing, the homogeneous $SiC_{0.7}$ layer entirely transforms into SiO_2, and then goes the oxidation of the interface "SiC film – Si substrate", in which the carbon concentration decreases uniformly with depth according to a Gaussian law. Thus, the concentration of carbon in the remaining layer begins to decrease. This leads to the appearance of phenomenon which is characteristic for the $SiC_{0.4}$, $SiC_{0.12}$, $SiC_{0.03}$ layers, namely, to shift of the minimum of the IR transmission peak up to 820 cm^{-1}, and to a decrease of the amplitude of the LO-phonon peak and their subsequent disappearance. Thus, size effects are confirmed, published by us in 2011.

3.3. Parameters Of Sic And C Films On Si Substrates Synthesized By Magnetron And Ion-Beam Sputtering

3.3.1. Parameters Of C Films On Si Substrates Synthesized By By Magnetron Sputtering

Carbon thin films were obtained by reactive magnetron sputtering using an ARC 2000 system. A graphite target with a diameter ~50 mm and a thickness of 3 mm was used. The magnetron sputtering mode parameters were: cathode voltage U_c = 470 V, the ion beam current I_{ion} = 35 mA and the argon pressure inside the chamber ~1 Pa. The carbon films were deposited on a set of cleaned silicon substrates. The temperature of the substrate was 75°C.

The presence of a sharp interface "C film - Si substrate" permits to investigate the thickness and density of the film by X-ray reflectometry (CompleXRay C6) by recording the angular dependence of the reflection coefficient using two spectral lines CuK_α (0.154 nm) and CuK_β (0.139 nm). The oscillations of intensity were observed, assigned to the interference of X-ray reflections from the boundaries of carbon layer (Fig. 24).

It is known that the density of graphite is 2.2 g/cm^3 and the density of diamond - 3.51 g/cm^3. Since the density of the resulting film was 3.32 g/cm^3 (Table 5 and Fig.24), we concluded that the diamond-like carbon film was synthesized.

Table 5. Determination of the density of the carbon layer by X-ray reflectometry and using the Henke program.

Layer	I_{max}	$I_{max}/2$	$2\theta_c$	θ_{cr} grade	θ_{cr} rad	ρ, g/cm³
C	962849	481425	0.529	0.2645	4.616	3.32

To determine the thickness, five narrow peaks of C, and a broad band of C were used (Fig. 24, table 5).

Table 6. Determination of the thickness of the layers in the system (C - C - Si) by X-ray reflectometry.

Layer	$(2\theta)_j$	$(2\theta)_i$	j - i	$2\theta_{av} = [(2\theta)_j-(2\theta)_i]/(j-i)$	λ, nm	$d = \lambda/2\theta$, nm
C	1.684	1.164	5	0.104	0.15405	84.9
C	2.732	2.080	1	0.652	0.15405	13.5

Simulation using the Henke program (http://henke.lbl.gov/ optical_ constants/) [24] allows to obtain a theoretical curve, which is close to the experimental (Fig.25). The main parameters of the layer system, which allow to obtain an acceptable agreement of experimental and theoretical curves, were:

1) the diamond-like carbon film of thickness d = 84 nm, density ρ = 3.3 g/cm³, and surface roughness σ = 1.5 nm;

2) a thin graphite layer of thickness d = 5 nm with the density ρ = 2.206 g/cm³, and the roughness of (C– C) interface σ = 0.13 nm;

3) the silicon substrate with density ρ = 2.33 g/cm³ and the roughness of interface (C–Si) σ = 0 nm.

Thus, diamond-like carbon film of thickness d = 84 nm, density ρ = 3.3 g/cm³, and surface roughness of σ = 1.5 nm on a silicon surface by magnetron sputtering was synthesized.

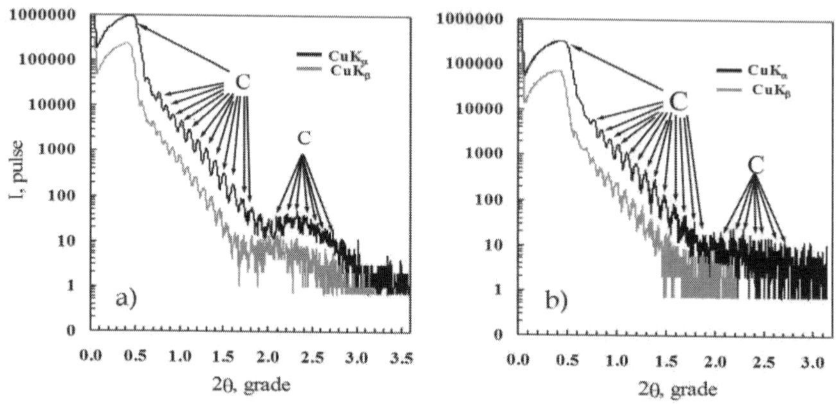

Figure 24. X-ray reflectometry using two spectral lines CuK_α (0.154 nm) and CuK_β (0.139 nm) (CompleXRay C6) of parameters of carbon films synthesized by magnetron sputtering: a) sample A; b) sample B.

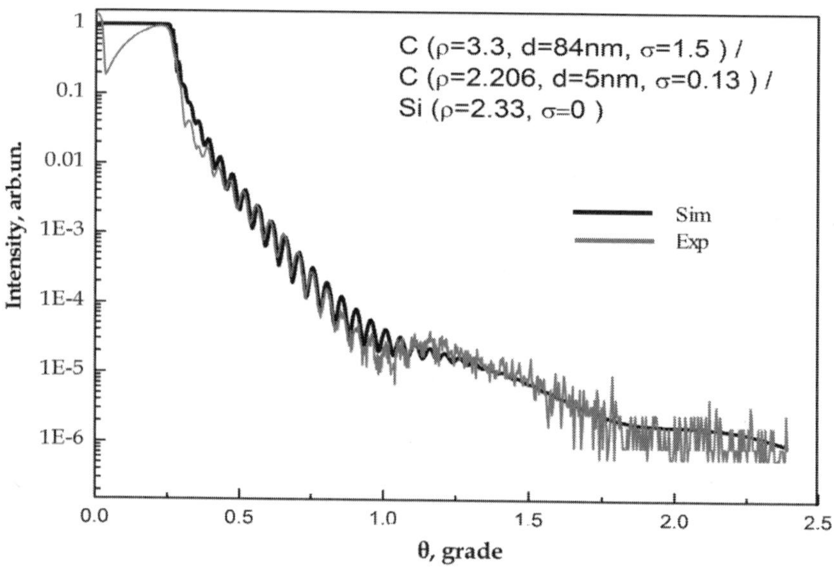

Figure 25. Simulation using the Henke program [24] of experimental results obtained by X-ray reflectometry of parameters of carbon films synthesized by magnetron sputtering.

3.3.2. Parameters Of Sic Films On Si Substrates Synthesized By By Ion-Beam Sputtering

SiC films were prepared by ion-beam sputtering. For the simultaneous deposition on silicon substrates of C and Si atoms, a two-component target consisting of the overlapping wafers of silicon and graphite was used. Sputtering of the target was carried out in an argon atmosphere. The formation of Ar ion beam was happening in the ring electrode system (a hollow cathode and an anode), and magnets with crossed electrical and magnetic fields. Discharge power was 108 W (2.7 kV, 40 mA), argon pressure in the chamber 5.9×10^{-2} Pa, substrate temperature - 20°C. Samples with SiC films were annealed at 1250°C in an argon atmosphere for 30 min.

The presence of a sharp interface "the SiC film - Si substrate" permits to investigate the thickness and density of the film by X-ray reflectometry ("ComplexXRay C6"). The oscillations of intensity were observed, assigned to the interference of X-ray reflections from the boundaries of silicon carbide layer (Fig. 26). It is known that the density of graphite is 2.2 g/cm^3, and of silicon - 2.33 g/cm^3, of silicon carbide - 3.2 g/cm^3 and of diamond - 3.51 g/cm^3. Since the density of the obtained film was 3.03 g/cm^3(Table 6 and Fig.26), we concluded, that the film close to silicon carbide, was synthesized. The film contains approximately [(3.03-2.33)/(3.2-2.33)] × 100% = 80.5% SiC, and [(3.2-3.03)/(3.2-2.33)] × 100% = 19.5% Si, i.e., about 80 atoms of C refer per 100 atoms of Si and the $SiC_{0.8}$ layer was formed.

Table 7. Determination of the density of the SiC layer by X-ray reflectometry and using the Henke program.

Layer	I_{max}	$I_{max}/2$	$2\theta_c$	θ_c, grade	θ_c, rad	ρ, g/cm^3
SiC	101255	50628	0.508	0.254	4.433	3.03

To determine the film thickness the position of maxima of five narrow peaks of SiC was used (Fig.26).

Simulation using the Henke program (http://henke.lbl.gov/ optical_ constants/) [24] allows obtaining a theoretical curve, which is close to the experimental (Fig.27). The main parameters of the layer system, which allow to obtain an acceptable agreement of experimental and theoretical curves, were:

1) the silicon carbide film of thickness d = 160 nm, the density ρ = 3.03 g/cm^3, and surface roughness of σ = 0.4 nm;

2) the silicon substrate with density ρ = 2.33 g/cm^3 and the roughness of interface (SiC-Si) σ = 1.5 nm.

Table 8. Determination of the thickness of the layer in the system (SiC – Si) by X-ray reflectometry.

Layer	$(2\theta)_j$	$(2\theta)_i$	j – i	$2\theta_{av} = [(2\theta)_j-(2\theta)_i]/(j-i)$	λ, nm	d = λ/2θ, nm
SiC	1.230	1.034	4	0.0490	0.15405	180.1

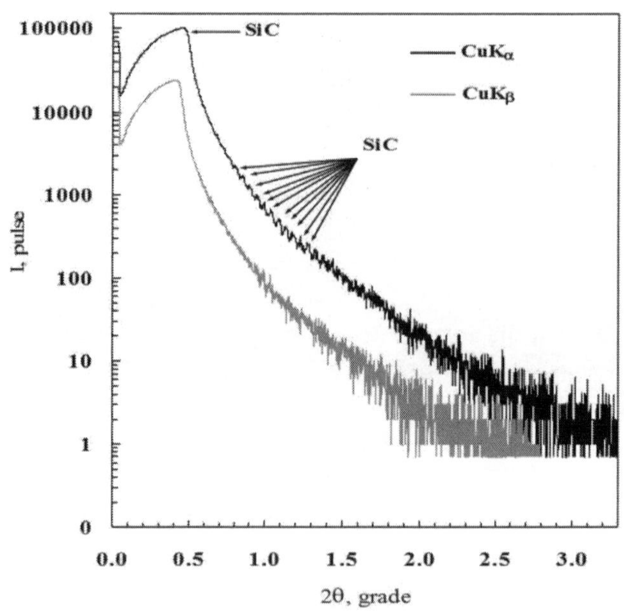

Figure 26. X-ray reflectometry using two spectral lines CuK_α (0.154 nm) and CuK_β (0.139 nm) (CompleXRay C6) parameters of silicon carbide films synthesized by ion-beam sputtering of the two-component target of silicon and graphite.

The film thickness d = 160 nm is different from the values of 180 nm, obtained from the average distance between the peaks $2\theta_{av}$. The reasons for the differences require further studies.

Thus, the silicon carbide film of thickness d = 160 nm, the density ρ = 3.03 g/cm^3, and surface roughness of σ = 0.4 nm on the silicon surface by ion-beam sputtering of the two-component target of silicon and graphite was synthesized.

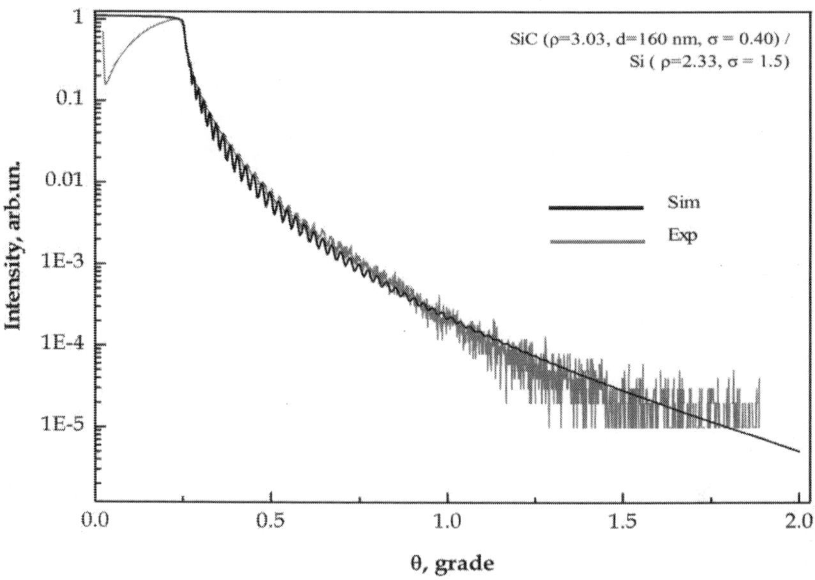

Figure 27. Simulation using the Henke program [24] of experimental results obtained by X-ray reflectometry of parameters of SiC films synthesized by ion-beam sputtering of the two-component target of silicon and graphite.

4. CONCLUSION

- For silicon layer with the Gaussian profile of implanted carbon atoms, the peak of longitudinal optical oscillations (LO-phonons) of SiC at 965-970 cm^{-1} is found and, it permits to calculate a number of optical parameters of film. The values of the low-frequency dielectric constant, ε_0, the effective charge, e*/e, and the dimensionless parameter, ρ, are equal to 9.82, 0.89, and 0.25, respectively. The dependences of LO-phonon peak amplitude, halfwidth and peak position from both an angle of incidence of the electromagnetic radiation on sample surface and the annealing temperature are determined. The difference between the LO-phonon curves behaviour indicates that the formation of SiC crystallites, i.e. the intensive formation of tetrahedral Si–C-bonds of necessary length and bond angles, in the case of (111) oriented substrate is not completed up to the silicon melting point, as for (100) oriented substrate that is completed at 1350°C.

- The temperature dependence of an area under contour of IR spectrum curve in wide temperature range 200 – 1400°C was used to obtain valueable new information about the structure transformation in the carbon implanted silicon layer. The high sensitivity of temperature dependence of area to a change of substrate orientation was found.

- A model of the ion implanted layer in the form of the system of the carbon-silicon clusters consisting of the carbon- and silicon atoms linked one with another by single-, double- and tripple bonds, and by single elongated-, sesqui, free ("dangling") and hybridized bonds, as well as resonances, is discussed. The infrared inactive flat nets and chains tied together by the infrared active clusters and, by tetrahedrally oriented bonds which are characteristic for silicon and silicon carbide, are described. Immediately after the implantation of carbon into the (100) and (111) oriented silicon at least 65% and 60% of carbon atoms are concentrated in these optically inactive clusters, respectively, if the implantation was carried out by a dose to obtain the stoichiometric concentration (E = 40 keV, D = 3.56×10^{17} cm^{-2}) of carbon and silicon. An over-barrier mechanism of the formation of Si and SiC crystallites is supposed and it explained by tendency of an isolated system to a minimum of free energy.

- For carbon implanted silicon layer (E = 40 keV, D = 3.56×10^{17} cm^{-2}) on (100) and (111) substrate the crystallization process in several temperature ranges was studied. For (100) oriented Si substrate it was shown:

    - 20-600°C – the formation of tetrahedrally oriented Si–C-bonds due to disintegration of clusters which consist mainly of the Si–Si, Si=Si and elongated Si–C-bonds;

- 600–800°C – the formation of Si crystallites and tetrahedrally oriented Si–C-bonds due to disintegration of strained Si-Si bonds and Si–C-bonds, which absorb at frequencies close to 700 and 750 cm^{-1}, recrystallization starts near the substrate and the surface and goes to the middle of layer;

- 800–1000°C – begins to dominate the absorption at 800 cm^{-1}, and near it, due to formation of tetrahedrally oriented bonds of the SiC crystallites, as well as bonds close to tetrahedral;

- 1000–1100°C - the growth of the crystallite size through consolidation of small crystallites of Si and SiC; decay of Si–C-bonds absorbing at 850 and 900 cm^{-1};

- 1100–1200°C – formation of new optically active clusters absorbing at frequencies of 750, 850 and 900 cm^{-1} due to the decay of optically inactive clusters;

- 1200–1250°C – the transformation of clusters absorbing at frequencies of 800-900 cm^{-1}, into clusters, which absorb at 700 and 800 cm^{-1}, and the growth of SiC crystallite size;

- 1250–1300°C – intensive decay of the tetrahedral Si–C-bonds, caused by the desorption of carbon and the destruction of small crystallites;

- 1300–1350°C – the decay of stable clusters with multiple bonds and the increase in the number and size of crystallites of SiC, and Si–C-bonds, close to tetrahedral;

- 1350–1400°C – a significant decrease of amount of the polycrystalline SiC due to desorption of carbon atoms.

- An influence of substrate orientation on β-SiC formation has been studied. It was shown that the SiC-synthesis in the temperature range 900–1000°C is more preferable on (100) oriented silicon substrates and at 1200–1300°C – on (111) oriented silicon substrates. In the case of the (100) oriented substrate the number of tetrahedrally oriented Si–C-bonds reached at 1000°C some maximum and does not change up to 1200°C, whereas in the case of orientation (111) the number of bonds increases smoothly in the range 900–1300°C.

- Using thermo-emf, it was determined a type of conduction of Si- and SiC-crystallites. The SiC-crystallites have p-type of conduction independently from the type of conduction of substrate, while the Si-crystallites have the same conduction as a substrate.

- During the prolonged high-temperature isothermal annealing (1200°C) a gradual decrease in the amplitude of the TO- and LO-phonon peaks of IR transmission, characteristic of ion-synthesized SiC, indicates the disintegration of the structure of homogeneous SiC film, ie the instability of these films at these temperatures.

- It is shown that the deformation of the rectangular Auger profile of C^{12} distribution in Si in comparison with the calculated profile, namely the thinning of the interface "the SiC film – Si substrate", and the increase of the carbon concentration at the surface and in regions near the maxima of the distributions for individual carbon ion energy (40, 20 keV), is caused by both the surface sputtering and the change in the composition of the silicon layer during high dose implantation of carbon.

- The presence of a sharp interface "the SiC film – Si substrate" permits to study the composition, the density and the thickness of the $SiC_{0.7}$ and $SiC_{0.95}$ films by X-ray reflectometry (CompleXRay C6) at small grazing angles θ by recording the angular dependence of the reflection coefficient using two spectral lines CuK_α (0.154 nm) and CuK_β (0.139 nm), although this method for ion-implanted layers usually does not apply. Using the Henke program for $SiC_{0.7}$ film was determined a surface layer with density 2.37 g/cm^3, which is close to cristobalite (SiO_2) 2.32 g/cm^3. Further located the film with a density 2.77 g/cm^3 and is close to quartz (SiO_2) 2.65 g/cm^3. A structure of thickness 65 nm and density ρ = 3.25 g/cm^3, is close to the silicon carbide - 3.2 g/cm^3. The study of $SiC_{0.95}$ film show the presence of surface layer with density 2.51 g/cm^3 (optical glass) and the silicon carbide layer of thickness 94 nm and density 3.06 g/cm^3.

- It is shown that the decomposition of optically active Si–C-bonds is more intense in the case of the (111) orientation of the silicon substrate. Based on the linear nature of reducing the number of Si–C-bonds in a homogeneous $SiC_{0.7}$ layer with increasing annealing time, it was concluded that the decay rate of silicon carbide does not depend on the depth of the oxidation front.

- It was confirmed the size effect caused by the small size of nanocrystals, which are manifested in the shift of the minimum of the peak of IR transmission up to to 820 cm^{-1}, the decrease of the amplitude of the LO-phonon peak and its subsequent disappearance during the oxidation of the interface “SiC film - Si substrate ”, where the concentration of carbon atoms decreases.

- Diamond-like carbon film of thickness d = 84 nm, density ρ = 3.3 g/cm^3, and surface roughness of σ = 1.5 nm on a silicon surface by magnetron sputtering was synthesized. The silicon carbide film of thickness d = 160 nm, the density ρ = 3.03 g/cm^3, and surface roughness of σ = 0.4 nm on the silicon surface by ion-beam sputtering of the two-component target of silicon and graphite was synthesized. Simulation using the Henke program allows obtain theoretical curves, which are close to the experimental curves.

REFERENCES

1. A. V. Afanas’ev, V. A. Il’in, A. V. Korlyakov, A. O. Lebedev, V. V. Luchinin, Yu. M. Tairov, 2011 Karbid kremniya. Vklad SPbGETU « LETI». Priznanie i perspektivy. V kn.: Fizika i tekhnologiya mikro- i nanosistem. S. 50-86. (Silicon carbide. The contribution of SPbGETU "LETI". Recognition and prospects. In the book. Physics and technology of micro-and nanosystems. P.50-86). Edited by Luchinin V.V. and Malinovski V.V. Publisher "Russian Collection". St. Petersburg 239 in Russian.

2. I. P. Akimchenko, K. V. Kisseleva, V. V. Krasnopevtsev, Yu. V. Milyutin, A. G. Touryanski, V. S. Vavilov, 1977a The structure of silicon carbide synthesized in diamond and silicon by ion implantation Radiation Effects 33 75 80

3. I. P. Akimchenko, Kh.R Kazdaev, V. V. Krasopevtsev, 1977b IK-pogloshenie β-SiC, sintezirovannogo pri ionnoi implantacii C v Si. (IR absorption of β-SiC, synthesized by ion implantation of C into Si). Fizika i Tekhnika Poluprovodnikov, T.11, vyp. 10 1964 1966 in Russian

4. I. P. Akimchenko, H. R. Kazdaev, I. A. Kamenskikh, V. V. Krasnopevtsev, 1979 Opticheskie i fotoelektricheskie svoistva struktury SiC–Si, poluchennoi pri implantacii ionov ugleroda v

kremnii. (Optical and photoelectric properties of the SiC-Si structure, obtained by implantation of carbon ions into silicon). Fizika i Tekhnika Poluprovodnikov, T.13, vyp.2 375 378 in Russian

5. I. P. Akimchenko, K. V. Kisseleva, V. V. Krasnopevtsev, A. G. Touryanski, V. S. Vavilov, 1980 Structure and optical properties of silicon implanted by high doses of 70 and 310 keV carbon ions Radiation Effects 48 7 12

6. P. A. Aleksandrov, E. K. Baranova, K. D. Demakov, F. F. Komarov, A. P. Novikov, S. Y. Shiryaev, 1986 Sintez monokristallicheskogo karbida kremniya s pomoshyu odnoshagovoi tehniki vysokointensivnogo ionnogo legirovaniya. (Synthesis of single-crystal silicon carbide by using one-step technique of high-intensity ion implantation). Fizika i Tekhnika Poluprovodnikov, T.20 1 149 152 in Russian

7. P. A. Aleksandrov, E. K. Baranova, A. E. Gorodetski, K. D. Demakov, O. G. Kutukova, S. G. Shemardov, 1988 Issledovanie raspredeleniya amorfnoi i kristallicheskoi fazy ionno-sintezirovannogo SiC v Si. (The investigation of the distribution of the amorphous and crystalline phases of ion-synthesized SiC in Si). Fizika i Tekhnika Poluprovodnikov, T.22 4 131 132 in Russian

8. E. K. Baranova, K. D. Demakov, K. B. Starinin, L. N. Strelcov, I. B. Haibullin, 1971 Issledovanie monokristallicheskih plenok SiC, poluchennykh pri bombardirovke ionami C+ monokristallov Si. (The study of single-crystal films of SiC, obtained by bombardment by C+ ions of Si single crystals). Doklady AN SSSR 200 869 870 in Russian

9. R. M. Bayazitov, I. B. Haibullin, R. I. Batalov, R. M. Nurutdinov, L. Kh Antonova, V. P. Aksenov, G. N. Mikhailova, 2003 Structure and photoluminescent properties of SiC layers on Si, synthesized by pulsed ion-beam treatment. Nucl. Instrum. and Meth. in Phys. Res., B. 206 984 988

10. A. I. Belov, A. N. Mihailov, D. E. Nikolichev, A. V. Boryakov, A. P. Sidorin, A. P. Grachev, A. V. Ershov, D. I. Tetelbaum, 2010 Formation and "white" photoluminescence of nanoclusters in SiOx films implanted with carbon ions. Semiconductors. 44 11 1450 1456

11. J. A. Borders, S. T. Picraux, W. Beezhold, 1971 Formation of SiC in silicon by ion implantation. Appl Phys. Lett 18 11 509 511

12. D. Chen, S. P. Wong, Sh. Yang, D. Mo, 2003 Composition, structure and optical properties of SiC buried layer formed by high dose carbon implantation into Si using metal vapor vacuum arc ion source. Thin Solid Films 426 1 7

13. P. Durupt, B. Canut, J. P. Gauthier, J. A. Roger, J. Pivot, 1980 RBS, Infrared and diffraction compared analysis of SiC Synthesis in C implanted silicon Mater. Res. Bull 15 1557 1565

14. F. L. Edelman, O. N. Kuznetsov, L. V. Lezheiko, E. V. Lubopytova, 1976 Formation of SiC and Si3N4 in silicon by ion implantation. Radiation Effects 29 13 15

15. A. Fissel, U. Kaizer, K. Pfennighaus, B. Schroter, W. Richter, 1996 Growth of 6H-SiC on 6H-SiC(0001) by migration enhanced epitaxy controlled to an atomic level using surface superstructures. Applied Physics Letters 68 1204 1206

16. A. Fissel, B. Schroter, U. Kaiser, W. Richter, 2000 Advances in the molecular-beam epitaxial growth of artificially layered heteropolytypic structures of SiC. Appl. Phys. Lett 77 2418 2420

17. N. Frangis, F. Nejim, P. L. F. Hemment, J. Stoemenos, J. Van Landuyt, 1996 Ion beam synthesis of β-SiC at 950°C and structural characterization. Nuclear Instruments and Methods in Phys. Res. B 112 325 329

18. N. Frangis, J. Stoemenos, J. Van Landuyt, F. Nejim, P. L. F. Hemment, 1997 The formation of 3C-SiC in crystalline Si by carbon implantation at 950°C and annealing- a structural study. Journal of Crystal Growth 181 218 228

19. V. C. George, A. Das, M. Roy, A. K. Dua, P. Raj, D. R. T. Zahn, 2002 Bias enhanced deposition of highly oriented β-SiC thin films using low pressure hot filament chemical vapour deposition technique. Thin Solid Films 419 114 117

20. N. N. Gerasimenko, O. N. Kuznetsov, L. V. Lezheyko, E. V. Lyubopytova, L. S. Smirnov, F. L. Edelman, 1974 Nekotorye svoistva plenok SiC, poluchennykh ionnym vnedreniem v strukture Ai-SiC-Si. (Some properties of the SiC films, obtained by ion implantation in the structure of Al-SiC-Si.). Mikroelektronika, T.3., Vyp.5 467 468 in Russian

21. J. F. Gibbons, W. S. Johnson, S. W. Hylroic, 1975 Projected Range Statistics. Part 1. 93p. 2nd ed. Dowden, Stroudsburg. PA.

22. N.L Glinka, 1985 Obshaya khimiya (General chemistry). 24-e izd. Pod red. Rabinovich V.A. L: Khimiya. 702 in Russian

23. S. K. Gupta, J. Akhtar, 2011 Thermal Oxidation of Silicon Carbide (SiC)- Experimentally Observed Facts. In the book "Silicon Carbide-Materials, Processing and Applications in Electronic Devices", edited by Moumita Mukherjee, InTech, Chapter 9. 207 230

24. B. L. Henke, E. M. Gullikson, J. C. Davis, 1993 X-ray interactions: photoabsorption, scattering, transmission, and reflection at E=50-30000 eV, Z=1-92. Atomic Data and Nuclear Data Tables. 2 181 342

25. S. Kerdiles, R. Rizk, F. Gourbilleau, A. Perez-Rodriguez, B. Garrido, O. Gonzalez-Varona, J. R. Morante, 2000 Low temperature direct growth of nanocrystalline silicon carbide films Mater.Sci.Eng., B 69 530 535

26. T. Kimura, Sh. Kagiyama, Sh. Yugo, 1981 Structure and annealing properties of silicon carbide thin layers formed by ion implantation of carbon ions in silicon. Thin Solid Films 319 327

27. T. Kimura, Sh. Kagiyama, Sh. Yugo, 1982 Characteristics of the synthesis of β-SiC by the implantation of carbon ions into silicon. Thin Solid Films 94 191 198

28. T. Kimura, Sh. Yugo, Zh. S. Bao, Y. Adachi, 1989 Formation of silicon carbide layers by the ion beam technique and their electrical properties. Nuclear Instruments and Methods in Phys.Res., B. 39 238 241

29. A. F. Khokhlov, D. A. Pavlov, A. I. Mashin, D. A. Khokhlov, 1994 Vozniknovenie dvoiinykh svyazyei kremnii-kremnii v plenkakh a-Si:H, obluchennykh nyeonom i uglerodom pri otzhige. (The appearance of double silicon-silicon bonds in the a-Si:H films, exposed by neon and carbon during annealing). Fizika i tekhnika poluprovodnikov. T.28, vyp.10. S. 1750 1754 in Russian

30. A. F. Khohlov, D. A. Pavlov, A. I. Mashin, Y. A. Mordvinova, 1987 Izovalentnoe legirovanie amorphnogo kremniya uglerodom. (Isovalent doping of amorphous silicon by carbon). Fizika I tehnika poluprovodnikov, T.21, vyp.3 531 535 in Russian

31. A. A. Lebedev, G. N. Mosina, I. P. Nikitina, N. S. Savkina, L. M. Sorokin, A. C. Tregubova, 2001 Investigation of the structure of (p)3C - SiC-(n)6H-SiC. Tech. Phys. Lett 27 1052 1054

32. A. A. Lebedev, A. M. Strel'chuk, N. S. Savkina, E. V. Bogdanova, A. S. Tregubova, A. N. Kuznetsov, L. M. Sorokin, 2002 Investigation of the p-3Cn-SiC/n+-6H-SiC heterostructures with modulated doping. Tech. Phys. Lett 28 12 1011 1014

33. A. A. Lebedev, D. K. Nelson, B. S. Razbirin, I. I. Saidashev, A. N. Kuznetsov, A. E. Cherenkov, 2005 Study of the properties of two-dimensional electron gas in p-3Cn-SiC/n+-6H-SiC heterostructures at low temperatures. Semiconductors 39 10 194 996

34. L. V. Lezhyeiko, Ye. V. Lyubopytova, 1976 Svoistva p-n i n-n-geteroperekhodov Si–β-SiC, poluchennykh ionnym vnedreniem. (Properties of p-n and n-n-heterojunctions Si-β-SiC, obtained by ion implantation). Fizika i tekhnika poluprovodnikov, T. 10 6 1749 in Russian

35. Y. Liangdeng, S. Intarasiri, T. Kamwanna, S. Singkarat, 2008 Ion beam synthesis and modification of silicon carbide. In book "Ion beam applications in surface and bulk modification of insulators".- IAEA, Austria, Vienna, IAEA-TECDOC-1607 63 92

36. F. Liao, S. L. Girshick, W. M. Mook, W. W. Gerberich, M. R. Zachariah, 2005 Superhard nanocrystalline silicon carbide films Appl. Phys. Lett 86 171913 171915

37. J. K. N. Lindner, 2003 High-dose carbon implantations into silicon: fundamental studies for new technological tricks Appl.Phys. A 77 27 38

38. R. H. Lyddane, R. G. Sachs, E. Teller, 1941 On the polar vibration of alkali halides. Phys.Rev 59 673 676

39. T. Morimoto, Y. Ogura, M. Kondo, T. Ueda, 1995 Multilayer coating for carbon-composites Carbon 33 4 351 357

40. D. M. Mukhamedshina, N. B. Beisenkhanov, 2012 Influence of Crystallization on the Properties of SnO2 Thin Films. In book: "Advances in Crystallization Processes". Chapter 9. InTech. Dr. Yitzhak Mastai (Ed.). 219 258

41. S. Nishino, C. Jacob, Y. Okui, 2002 Lateral over-growth of 3C-SiC on patterned Si(111) substrates. Journal of Crystal Growth, 237-239 1250 1253

42. K. Oguri, T. Sekigawa, 2004 Heat resistant material and hot structure member both space shuttle, space shuttle, and method for producing heat resistant material for space shuttle. United State Patent Sep.16 US2004/0180242 A1

43. T. Pajagopalan, X. Wang, B. Lanthouh, C. Ramkumar, 2003 Low temperature deposition of nanocrystalline silicon carbide films by plasma enhanced chemical vapor deposition and their structural and optical characterization. J. Appl. Phys 94 3 5252 5260

44. L. Pauling, P. Pauling, 1975 Chemistry. Freeman W.H. and company. San Francisco

45. R. E. Peierls, 1956 Quantum theory of solids Oxford: Clarendon Press 54 58

46. G. Raabe, J. Michl, 1985 Multiple Bonding to Silicon Chem. Rev. 85 5 419 509

47. W. Rothemund, C. R. Fritzsche, 1974 Optical absorption and electrical conductivity of SiC films produced by ion implantation J.Electrochem. Soc.: Solid-state science and technology 121 4 586 588

48. N. S. Savkina, A. A. Lebedev, D. V. Davydov, A. M. Strelchuk, A. S. Tregubova, C. Raynaud, J. P. Chante, M. L. Locatelli, D. Planson, J. Nilan, P. Godignon, F. J. Campos, N. Nestres, J. Pascual, G. Brezeanu, M. Batila, 2000 Low-doped 6H-SiC n-type epilayers grown by sublimation epitaxy Materials Science and Engineering B77 50 54

49. A. V. Semenov, V. M. Puzikov, M. V. Dobrotvorskaya, A. G. Fedorov, A. V. Lopin, 2008 Nanocrystalline SiC films prepared by direct deposition of carbon and silicon ions Thin Solid Films 516 10 2899 2903

50. A. V. Semenov, V. M. Puzikov, E. P. Golubova, V. N. Baumer, M. V. Dobrotvorskaya, 2009 Low-temperature production of silicon carbide films of different polytypes Semiconductors 43 5 685 689

51. A. V. Semenov, A. V. Lopin, V. M. Puzikov, V. N. Baumer, I. N. Dmitruk, 2010 Fabrication of heterostructures based on layered nanocrystalline silicon carbide polytypes Semiconductors 44 6 816 823

52. C. Serre, A. Romano-Rodríguez, A. Pérez-Rodríguez, J. R. Morante, L. Fonseca, M. C. Acero, R. Kögler, W. Skorupa, 1999 SiC on SiO2 formed by ion implantation and bonding for micromechanics applications. Sensors and Actuators A: Physical 74 1-3 169 173

53. H. Spillman, P. R. Wilmott, 2000 Growth of SiNx and SiCx thin films by pulsed reactive crossed-beam laser ablation. Applied Physics. A.-Materials Science & Processing 70 3 323 327

54. W. G. Spitzer, D. Kleinman, D. Walsh, 1959 Infrared properties of hexagonal silicon carbide Phys. Rev. 113 1 127 132

55. K. Srikanth, M. Chu, S. Ashok, N. Nguyen, K. Vedam, 1988 High-dose carbon ion implantation studies in silicon Thin Solid Films 323 329

56. Y. Sun, T. J. Miyasato, 1998 Characterization of cubic SiC films grown on thermally oxidized Si substrate. J. Appl. Phys 84 5 2602 2611

57. D. I. Tetelbaum, A. N. Mikhaylov, A. I. Belov, V. K. Vasiliev, A. I. Kovalev, D. L. Wainshtein, Y. Golan, A. Osherov, 2009 Luminescence and structure of nanosized inclusions formed in SiO2 layers under double implantation of silicon and carbon ions. Journal of Surface Investigation. X-ray, Synchrotron and Neutron Techniques 3 5 702 708

58. A. G. Touriyanski, S. A. Aprelov, N. N. Gerasimenko, I. V. Pirshin, V. M. Senkov, 2007 Otnositelnaya rentgenovskaya reflektometriya diskretnykh sloistykh struktur (Relative X-ray reflectometry of discrete layered structures). Pis'ma v ZHTF (Technical Physics Letters)33 5 87 94

59. A. G. Touriyanski, N. N. Gerasimenko, S. A. Aprelov, I. V. Pirshin, A. I. Poprygo, V. M. Senkov, 2005 Investigation of ion-implanted layers by X-ray reflectometry. Proc. SPIE. X-Ray and Neutron Capillary Optics II. 5943 143 149

60. A. Touryanski, N. Gerasimenko, I. Pirshin, V. Senkov, 2009 Mnogofunktsionalnyi rentgenovskii reflektometr dlya issledovaniya

nanostruktur (Multipurpose X-ray reflectometer for the study of nanostructures). Nanoindustriya (Nanoindustry) 5 40 45 in Russian

61. A. G. Touryanski, A. V. Vinogradov, I. V. Pirshin, 2000 X-ray reflectometer. Patent no.6041098,US Cl. 378-70. Official Gazette March 21,2000 2960

62. A. G. Touryanski, A. V. Vinogradov, I. V. Pirshin, 2000 Two-channel X-ray reflectometer Nucl. Instrum. and Meth. in Phys. Res. A 448 184 187

63. H. Yan, B. Wang, X.M. Song, L. W. Tan, S. J. Zhang, G. H. Chen, S. P. Wong, R. W. M. Kwok, W. M. L. Leo, 2000 Study on SiC layers synthesized with carbon ion beam at low substrate temperature. Diamond and related materials 9 1795 1798

64. R. West, M. J. Fink, J. Michl, 1981 Tetramesityldisilene, a Stable Compound Containing a Silicon-Silicon Double Bond. Science. 214 1343 1344

65. J. S. Williams, J. M. Poate, 1984 Ion Implantation and Beam Processing. Academic Press, Melbourne, 1984. Edited by Williams J.S., Poate J.M. 419 p. 30 31

66. J. M. Ziman, 1960 Electrons and phonons. Oxford: Clarendon Press. 209

2

TOWARD LABEL-FREE BIOSENSING WITH SILICON CARBIDE: A REVIEW

Oren Cooper[1], Bei Wang[2,3], Christopher L. Brown[2], Joe Tiralongo[1], and Francesca Iacopi[2,3], (Member, Ieee)

[1]Glycomics Institute, Grif_th University, Southport, QLD 4222, Australia

[2]Queensland Micro and Nanotechnology Centre, Grif_th University, Nathan, QLD 4111, Australia

[3]Environmental Futures Research Institute, Grif_th University, Nathan, QLD 4111, Australia

ABSTRACT

Recent innovation in microelectrical–mechanical systems (MEMSs) and plasmonics-based technologies has opened up perspectives for label-free sensing of biological and chemical analytes. Label-free sensing would enable increased sensitivity and miniaturization capabilities for biosensing devices. Silicon carbide is a semiconductor material that happens to possess ideal properties for augmenting both the MEMS/nanoelectromechanical systems and the plasmonics routes. It has remarkable chemical and biological inertness resulting in a high degree of biocompatibility, as well as pronounced mechanical resilience. In addition, it is an efficient (low loss) plasmonic metamaterial. Its cubic polytype can be grown on silicon wafers, allowing easy micromachining into building blocks for sensing devices, scalable to large volume production. Finally, silicon carbide is an ideal starting material for a controlled, wafer-scale growth of graphene, offering an additional wealth of excellent properties for nanosensing. The combination of all of these capabilities makes silicon carbide an outstanding material platform for the realization of label-free, analyte-specific, and highly sensitive biochemical molecule detection systems. These technologies will open exciting horizons in terms of high throughput, efficient drug screening, and early pathogen detection.

SECTION I

INTRODUCTION

The development of label free, highly sensitive and molecule specific analyte detection technologies is of high relevance for a number of applications in the bio-medical area, including medical diagnostics and drug screening. Therefore, the availability of a robust, bio-compatible functional material which can be structured into a miniaturized sensing device is highly desirable. Such a system would enable highly efficient, high throughput chemical analysis and potentially minimally invasive endoscopic sensing.Silicon carbide is a wide band-gap semiconductor [1] [2] [3], with demonstrated bio- and hemo- compatibility [4] [5] [6] [7] [8]. Also, similar to silicon, the silicon carbide surface can be functionalized to capture or bind with specific reagents, and its use has been demonstrated in neural probes, enabling among others applications such as brain-machine interfaces [9]. Moreover, silicon carbide is a suitable template for the direct growth of epitaxial graphene, likely also a bio-compatible material which can bring exciting additional functionalities and capabilities to the silicon carbide system, such as low-loss plasmonic technologies [10].

This paper will first present a review of the status and recent progress in the area of label free detection technologies. We will focus on optical (plasmonics) and microelectrical-mechanical (MEMS) based sensing, as potentially the most accurate and molecule-specific sensing technologies, indicating advantages and limitations. A general introduction on silicon carbide will follow, focusing on how this wide band-gap semiconductor, as well as its combination with graphene, can further this field and open exciting horizons for bio-medical applications.

SECTION II

LABEL FREE DETECTION

Current established DNA [11], protein [12], [13], glycan [14], [15] and lectin [16], [17] arrays allow for high-throughput multiplexing with reduced sample volume. However, without appropriate quantitative controls and complex algorithms [18], results remain qualitative due to the requirement for fluorescent, photochemical or radioisotope tagging [19]. These labels and their appropriate laser scanners are commercially available, however, background interference due to the label itself and the need for adequate signal controls are serious limitations to this technology. Labelling probe molecules for large-scale studies is tedious, expensive and limited by various factors.For instance size and position of the label can induce conformational strains and steric hindrance, affecting the probes ability to interact with target structures [20]. These factors become exacerbated when probe molecules (metabolites, oligonucleotides, peptides and small organic molecules) are smaller than the label being used. Non-specific binding to the array platform will also contribute to the intrinsic fluorescence being measured and can give rise to false positives. Finally the extent of fluorescent labelling also needs to be optimized for various probe molecules to ensure normalisation between results. Using the established array technology in combination with real time label free detection systems would enable a paradigm shift within molecular and structural biology.

Accordingly, advances in label free detection methods are highly desirable for next generation bio-sensing platforms due to their potentials for increased sensitivity and direct measurement [21].

A. Plasmonic Sensors

From the discovery of surface plasmons in 1968 [22], plasmonic materials have emerged offering next generation label free detection. Plasmonic materials possesses a negative real and small positive imaginary dielectic constant, capable of supporting a single guided mode of electromagnetic field [21], [23]. Here the energy and momentum of a photon is coupled to a free electron gas in the form of surface plasmons. The surface plasmon is transversely magnetic; as such the vector of the magnetic field lies in plane of the metal-dielectic interface and is perpendicular to the direction of propagation. This is possibly due to surface plasmon sensitivity to the refractive index changes around metallic structures by electromagnetic radiation at a metal-dielectric interface [23], [24].

Plasmon based resonance sensing can be broadly separated into two types; the first is propagating surface plasmon resonances (PSPRs) that rely on evanescent electromagnetic waves bound by planar metal-dielectric interfaces. The second type is localized surface plasmon resonances (LSPRs), where electromagnetic waves are confined on metallic nanostructures [25]. Unlike conventional optical sensors that rely on labels (fluorophore and chromophore), surface plasmon resonance can transduce the binding event due to changes in the local refractive index when the target analyte binds to the surface.

Since the realization in 1980 that surface plasmon resonance (SPR) could be an outstanding probe of surface chemistry [26], this technique has been extensively used to elucidate binding kinetics, conformational changes and quantifications of chemicals; small ions and biomolecules immobilized to the surface [27], [28]. Use of thin metallic film allows plasmons to propagate hundreds of micrometers along the metal surface with an associated electric field that decays exponentially allowing for detection of refractive index through intensity, wavelength or angle shifts. Enhancements in this field have seen detection below the submicrometer range from improved molecular adsorptions [29] [30] [31], enlargement of binding site [32], [33] and introduction of signal enhancers on the metal surface [34] [35] [36]. SPR is a versatile tool as the monitored optical parameter can vary based on user specification including; measurement of the wavelength where resonant coupling occurs; the phase of the light; and the intensity of binding based on measurement of the shift in the angle at which resonant coupling takes place [37], [38]. Conventional SPR has long been the technology of choice for label free detection [39], however it does not match the demands of current microarray technology; being unable to couple the high throughput platform with sensitivity [40]. Conventional SPR is also significantly expensive as multiplexing relies on only a few flow cells limiting analysis to 50 spots [41]. Alternatively LSPR offers a cheaper alternative for smaller laboratories due to integration of lab-on-chip (LOC) technologies [42]. LOC technologies allow for cheaper purchase of one-time assays capable of producing vast quantitative bioinformation.

LSPR confines excited electromagnetic waves onto metallic nanostructures and by controlling physicochemical properties the spectral position and magnitude can be altered. Optimization of size, shape, composition, interstructural spacing and local dielectric environment have led to the advent of gold nanorings [43] [44] [45], metallic nanoislands [46], [47] and nanoholes in thin gold films [31], [48]. Unlike PSPRs, these nanostructures show enhanced sensitivity, as there is less interference from the bulk refractive index [32], [49], [50]. In addition to the refractive sensitive capability of LSPR, electromagnetic field enhancements generated around them has been shown to be ideal when coupled with surfaced enhanced Raman scattering (SERS). Nanosphere lithography (NSL) used to develop nanoparticle arrays (most commonly triangular structures) with tuneable LSPR has been previously developed as a self-assembled monolayer [34], [36]. Both experimental and theoretical studies have demonstrated that the sharp tip nanotriangles create an electromagnetic enhancement factor as large as [37], [38]. Further advances have seen atomic layer deposition with alumina overlayer to particles fabricated via nanosphere lithography, resulting in significant increases in thermal stability [43], [51]. The electromagnetic mechanism of SERS does not require direct contact of the surface due to the electric field extending within a few nanometers of the surface. Enhancement of Raman scattering can be exploited to measure systems that require surface-immobilized biological molecules [46], [52] however research is still required to enhance the multiplexing capabilities of this platform.

However, the current metal-based plasmonics sensing technologies are strongly limited by high resistive losses, which have restricted the development of these devices.Approaches to minimize metallic losses through discovery of better plasmonic materials offer promising alternatives, via doping, alloying and careful band structure engineering [53], [54]. As such low-loss plasmonic metamaterials have emerged with longer plasmon lifetimes [55] [56] [57] [58]. This category of low–loss materials include silicon carbide and graphene as both are tunable and have carrier concentrations high enough to provide a negative real permittivity [55].

B. MEMS Cantilever Sensors

Micro-electro-mechanical systems (MEMS) offer an alternative approach for quantitative molecular recognition. These systems rely on silicon and wide-bandgap (WBG) semiconductors that allow for increased stability, biocompatibility and further miniaturization for resonant nanoelectromechanical systems (NEMS) [59], [60]. The ability to detect multiple target molecules in small sample volumes has maintained a significant relevance for research into early detection of disease, and therefore platforms that satisfy these requisites are appealing. Microcantilevers arrays have emerged as a very promising candidate for label free multi-target detection that is both sensitive and selective in small volumes of sample. MEMS cantilever sensors are another next generation platform readily fabricated on silicon wafers [48], [61]. An array can be fabricated through a "top-down" approach to release spring microbeams. These cantilevers typically measure approximately 50- long, 10- wide and 0.3- thick and respond to surface stress variation from chemical or biological process [50], [62]. Cantilevers can be designed with very small force constants (0.008 - 30 N/m) making them extremely sensitive to variations in adsorption of molecules [63], [64].

Capture of a molecule onto the surface of a cantilever can be measured thanks to the deflection of the cantilever from adsorption-induced forces. In a dynamic regime, mass adsorption on the surface of a microbeam can result in an observable shift from its natural resonance frequency, as a mass change will affect the spring constant of the system [65]. Both adsorption induced cantilever deflections and frequency shifts can be monitored simultaneously [51]. Initially cantilevers were not considered very promising mass sensors due to non-uniformities in non-specific binding over the entire length of the cantilever resulting in variations in spring constant. However, by designing cantilevers with a localized adsorption area at the terminal end of the cantilever the contribution of the differential surface stress can be entirely attributed to mass loading [66]. The resonance frequency can shift due to changes in mass and spring constant.Surface area has been shown to increase sensitivity of mass detection, leading to nanopatterning and nanofabricated holes for increasing the total absorbed mass [61].Despite being well suited for mass detection in air and vacuum, detection of absorbed mass under solution is best detected through surface stress variations; due to poor resolution of cantilever resonance frequency in liquid environments [62]. In general deflection of cantilever can be divided into two different modes for detection; static and dynamic.

1) Cantilever Deflection-Based Sensing

a: Static Detection

Static mode of deflection is a result of variations in surface stress on the opposing surface of the microcantilever. Surface stress variations exist as changes in surface energy density or tension, occurring through molecule adsorption that decreases surface free energy. By restricting adsorption to one side of a cantilever, a differential surface stress between each side of the beam is produced, leading to cantilever bending.This surface stress can be explained by Stoney's formula [67], which shows that the longer the cantilever the more sensitive it becomes to surface stresses *(see figure 1)*. The sensitivity also relies on the detection technique. Optical detection techniques are most commonly used; however they are limited due to a narrow dynamic range and parasitic deflection [68]. As such the dynamic mode of detection is more beneficial being insensitive to the drift of the deflection signal and increasing reproducibility.

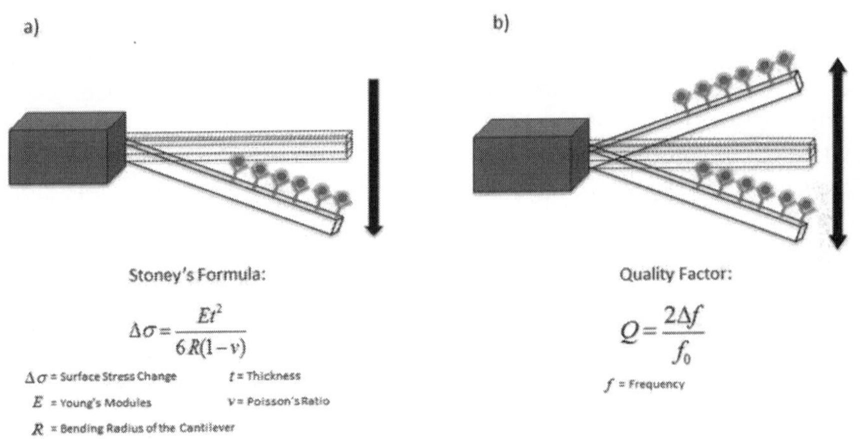

Figure 1. Schematic of the two different cantilever deflection based sensing. (a) Static detection through changes in surface stress. (b) Dynamic detection through changes resonance frequency.

b: Dynamic Detection

The dynamic mode of detection is directed at the variation in vibration frequency of the beam due to specific adsorption [69]. Unlike static measurements of surface stress, cantilever deflections leads to a "dynamic" action that can be used to quantitatively detect mass loading. Microcantilever beams freely resonant at their natural frequency due to thermal excitation. Mass loading leads to a decrease in vibrational frequency allowing for direct measurement by plotting the displacement amplitude against frequency [70] (*see figure 1 b*). Each resonance mode has its own quality factor that indicates the sharpness of a resonant peak [71]. Q factors are liable to damping effects from both liquid environments and geometry of the cantilever which can lead to poor frequency resolution [72], [73]. Higher factors allow for lower minimum detectable resonance shifts.

c: Read Out Mechanisms

In the last decade many techniques to measure cantilever resonance or deflection, have been demonstrated including; optical beam deflection [63], [74], [75] piezoresistivity [21], [76] [77] [78]; piezoelectricity [27], [28]; electron tunneling [79], [80] and capacitance [81]. Optical beam deflection was the first reported technique to exploit the mechanics of cantilever deflection by measuring bending by reflecting a focused beam of light from the tip of the cantilever into a position-sensitive detector. Optical beam deflection is a popular method for detection due to its sensitivity to capture cantilever bending in the sub-nanometer range [82] [83] [84]. A beam from a solid-state laser diode is focused onto the apex of the cantilever leading to a reflection that is captured by a positive sensitive detector of two closely spaced photodiodes. A differential amplifier captures this output signal and the measured distance traveled by the reflected laser beam is calculated proportionally to cantilever bending. Unfortunately this method relies on a complex design system, which can prove expensive to manufacture.

Piezoresistivity is an effect that semiconductors such as silicon and silicon carbide show as an intrinsic resistance change sensitively as a function of bending. A piezoresistive detection approach, measures the resistance of the material, which varies as a function of applied stress. The length of a piezoresistive cantilever again dictates resistance with a range typically between 1-5 $k\Omega$. Ramussen *et al.*, (2003) developed multilayer cantilever beams with single crystal silicon as the active functional element to detect variations in resistance as a function of deflection. Silicon doping of the beams was restricted to the neutral axis and the active silicon sandwiched between an insulating layer of silicon nitride and silicon dioxide. Cantilevers fabricated to this design with an insulating layer enable highly sensitive piezoresistive detection in liquid environments [85]. Increases in sensitivity have also been achieved by reducing the thickness of a cantilever. Piezoelectrical detection is possible by coating cantilevers with piezoelectric materials to generate a measurable charge in response to cantilever bending [86], [87]. A drawback to this technique is the extensive fabrication due to encapsulation of the electrically active parts.

Lee *et al.*, (1995) devised a self-excited piezoelectric cantilever by sandwiching a thin piezoelectric film of zinc oxide between two aluminum layers. Gravimetric testing revealed high sensitivity for the detection of chemical vapor concentration and relative humidity using an acoustic output transducer [88]. Using millimeter sized cantilever sensors, Campbell & Mutharasan (2006) demonstrated detection of *Bacillus anthracis*spores in liquid medium under both stagnant and flow conditions. These cantilevers were very sensitive with detection down to 300 spores/mL, as well as highly selective for*B. anthracis* in a mixture with *Bacillus thuringiensis* spores at ratios up to 1:500 [86]. Another work by Hwang *et al.*, (2004) fabricated self-sensing piezoelectric cantilevers for detection of prostate specific antigen (PSA). Using a PSA antibody immobilized to a calixcrown self-assembled monolayer, they were able to show the resonance frequency shift of the cantilever was proportional PSA concentration [68]. This group also employed the same fabrication technique using an immobilized C reactive protein antibody to detect C-reactive protein [89].

C. Bottom-Up Chemical Nanosensors With Nanotubes and Nanowires

Nanotubes and nanowires are one-dimensional nanostructures that have attracted increased interest as NEMS elements fabricated using a "bottom-up" approach. These structures contain excellent intrinsic properties and a low dimensionality. Wagner & Ellis (1965) first discovered nanowires (NW) through vapor-liquid-solid (VLS) mechanism by depositing semiconductor atoms under a supersaturated liquid catalyst to grow an epitaxial wire [90]. Since this innovation, the growth of many different NWs using VLS has been fabricated using various semiconductor atoms such as; Si, SiC, GaAs, Ge, and GaN [91] [92] [93] [94]. A study by Cui Y *et al.*, (2001) developed boron-doped silicon nanowires (SiNWs) to create a highly sensitive, real-time electrically based sensor [95]. In 2004 Patolsky *et al.*, using this same principal was able to show binding of a single virus particle [96]. The nanowire sensing approach is advantageous over conventional SPR in terms of sensor packing densities, demonstrating the ability to incorporate 2400 nanowire sensors onto one array [97]. This allows for reduced sample volume coupled with a higher sensitivity and high throughput function. Gamby *et al.*, (2009) fabricated gold NW (170 nm in diameter) in a polycarbonate microchannel through electro-crystallization techniques. These NWs demonstrated a measurable increase in Raman cross-section and could serve as a SERS active dielectric sensor easily integrated into lab-on-chip systems [52]. Nanowires are also suitable biological materials as they don't display acute (100 hr) toxicity towards cells, in fact, cells are capable of degrading NW into aggregates within days [98].

Carbon nanotubes (CNTs) are hollow variants of solid NW structures; first discovered by Iijma (1991), CNTs initially generated interest in the field of nanoelectronics due to their superconductivity capabilities [99]. Since their advent nanotubes have undergone extensive theoretical and experimental application to tailor desired growth and electronic properties [100] [101] [102]. Similar to MEMS structures nanotubes have comparable thermal and electrical conductivity and can be grown through CVD. CNT electrical properties can be controlled through chirality and number of carbon layers [42], [103]. Studies have demonstrated CNTs are useful label free immune sensors; for instance Okuno *et al.*, (2006) fabricated a single walled carbon nanotube (SWNTs) array to detect total PSA using differential pulse voltammetry. These single walled carbon nanotubes increased electron transfer improving the limit of detection to 0.25 ng/mL [100]. Silicon carbide nanotubes (SiCNTs) were first synthesized 10 years later by Pham-Huu *et al.*, (2001) and often advantageous over CNTs, particularly with respect to stability at high temperature, ease for sidewall decorations and semiconducting potentials irrespective of chirality [103]. SiCNTs have been shown to be exceptional in the detection of harmful gases, including; CO, NO [104], [105] and HCN [103]. Using SiC for fabrication of SiCNTs introduces interesting magnetic and electrical properties which can be manipulated by varying the surface decoration with SiH_3 and CH_3 functionalities [102].

D. Biological Applications

Microcantilever LOC platforms capable of detecting antigen-antibody interaction [50], protein-protein binding [106], DNA hybridization [64], [74] and DNA-protein interaction [83] have already been fabricated. Microcantilevers offer a comparable bimolecular detection platform to perform such multiplexing and label free analysis of these biomolecules with unparalleled sensitivities. As cantilever bending is sensitive enough to register the free energy change induced from binding events; as such immobilization of antibody molecules, leads to measurable cantilever deflection. For example ssDNA can be immobilized on the gold side of an asymmetrically doped cantilever using a thiol linker. This adsorption of ssDNA results in a surface stress variation between 30-50 mN/m [64], [74]. This ssDNA can be used as a probe to detect complementary sequences and would serve as an adequate platform to detect mutations such as single nucleotide polymorphisms [83], [107].

Fritz *et al.*, (2000) demonstrated optical beam deflection technique to detect nucleic acid hybridization on gold-coated silicon cantilevers. A 5-thio modified synthetic DNA oligonucleotides with different base sequences were immobilized and later detected in liquid through surface stress shifts of asymmetrical doping leading to cantilever deflection. In this same work it was shown that these cantilevers were able to distinguish between complementary oligonucleotides and a pair with single base mismatch between the DNA sequences being detected [74]. Advances on this principle led to the monitoring of restriction and ligation of cantilever coated with DNA. Stevenson *et al.*, (2002) functionalized silicon cantilevers coated in gold with 3-aminopropyltriethoxysilane (APTES) to monitor restriction and ligation of DNA. An oligonucleotide with the *Hind*III restriction site was immobilized to the APTES monolayer, followed by digestion with *Hind*III. This scission led to cantilever negative deflection due to the shortened oligo with a single stranded sticky end. Subsequent hybridization with a suitable second oligo in the presence of ligase and thus extension of immobilized DNA led to a commensurate cantilever deflection in the opposite (or positive) direction due to increases in mass loading [107].

Ilic *et al.*, (2000) was the first to report a high sensitivity detection of *Escherichia coli*using antibody layer coated silicon nitride cantilevers. The resonant frequency shifts measured as a function of mass loading from cell depositions was correlated with the number of cells bound on the surface [108]. Building on this work, Ilic *et al.*, (2004) were also the first to demonstrate detection of Baculovirus particles bound selectively to an AcV1 antibody monolayer immobilized onto cantilevers. These piezoelectric cantilevers allowed for single virus particle detection by measuring the resonant frequency shift [109]. Gfeller *et al.*, (2005) realized the cantilever biosensing potential for rapid real time detection of label free *E. coli* growth by coating cantilevers in nutritive agarose with bacteria. Resonance frequency shifts due to increasing mass were matched to the conventional bacteria growth curves, sensitive enough to observe all characteristic phases. A high sensitivity of 50 pg/Hz was calculated to be approximately 100 *E. coli* cells. These cantilevers were also shown to augment the ability to rapidly assess antibiotic resistance by either incorporating or omitting antibiotic coating on the cantilever [110].

Nugaeva *et al.*, (2004) using silicon cantilevers coated with 30 nm of gold functionalized with concanvalin A, fibronectin or immunoglobulin G demonstrated spore immobilization and germination of mycelial fungus *Aspergillus niger (see figure 2)*. Shifts in resonance frequency could be detected within a few hours as opposed to several days. Biosensing capabilities of these cantilevers detected fungi in the range of $10^3 - 10^6$ CFU/mL, and thus would be suitable for use in medical and agricultural diagnostic for both food and water quality monitoring [111]. Savran *et al.*, (2004) immobilized anti-Taq aptamer to silicon nitride cantilevers to investigate Taq polymerase binding affinity. Various concentrations of Taq polymerase (0.3-500 pm) was applied to the cantilevers resulting in polymerase aptamer binding and subsequent induce surface stress and deflection (3-32 nm depending on concentration) [112]. The resulting curve was fit using the least-squares method to reveal a K_d of $\sim$15 pM which was comparable to previous reports performed in solution [113]. Glucose detection has also been performed with glucose oxidase coated cantilevers [114] [115] [116]. These examples emphasize the high sensitivity and high throughput capabilities of microcantilever array platforms. The principles of surface plasmon resonance applied to these microstructures realize the potential of real time miniaturized multiplexing assays.

SECTION III

SILICON CARBIDE

A. Basic Properties & Polytypes

Moissianite, an extremely rare crystal formation of silicon carbide (SiC), is typically found in minute quantities in corundum deposits, kimberlitic and meteorites. SiC is a tetrahedron of four carbon atoms covalently bound to a centre silicon atom resulting in a 2-dimensional polymorphism. The high mechanical and chemical stability of SiC can be attributed to very short bond lengths, as each carbon is located 3.08 Å from each other and 1.89 Å from the silicon which leads to a very high bond strength [49], [117] [118] [119]. Si and C can form a bilayer introducing a double layer-stacking variable that defines over 200 different polytypes of SiC. These polytypes have been characterized based on the SiC bilayers that can assume a different hexagonal frame lattice and stacking sequence (*see figure 3*). 4H- and 6H- SiC are hexagonal polytypes based on the 4 and 6 respective bilayers required for the basic structure; alternatively, 3C-SiC (-SiC) is an example of a cubic polymorphic structure comprised of 3 bilayers [120], [121].

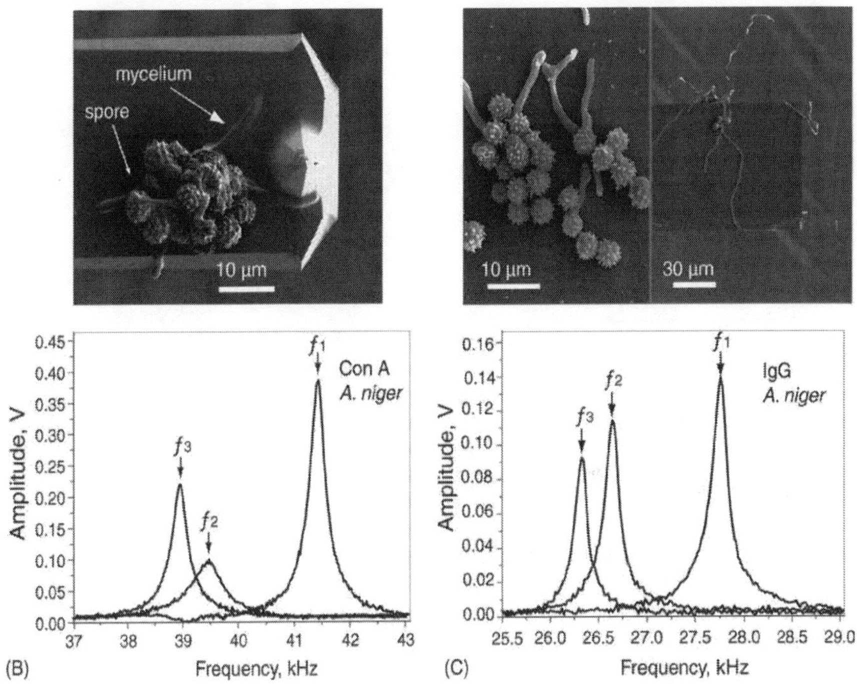

Figure 2. Scanning electron microscopy images of immobilized fungal spores and corresponding frequency spectra of the cantilevers after spore immobilization and spore growth. (A) Mycelial fungus *Aspergillus niger* on Con A coating; mycelia growing from spores after 12 hours (B) *A. niger*on IgG coating, active mycelial growth from spores (left image–growth time 4 h, right image–growth time 9 h). Resonance frequencies: –unloaded cantilever, –cantilever with immobilized spores,–cantilever with growing immobilized spores after exposure. The first resonance frequency shift corresponds to fungal spore mass loaded on the cantilever; second shift corresponds to *A. niger* mycelial growth. Reprinted with permission from [107].

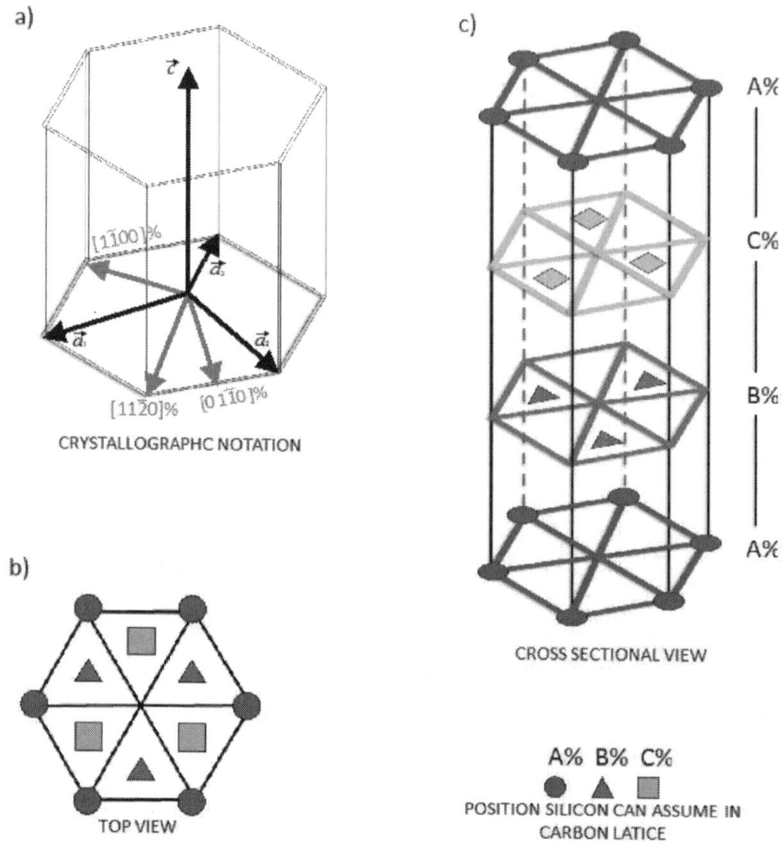

Figure 3. (a) Hexagonal crystallographic notation, which is used for SiC crystals independent of the actual lattice symmetry. It is based on four Miller-Bravais indices a_1, a_2, a_3 and c (shown in black) where: $a_1 + a_2 + a_3 = 0$. The grey vectors denote the different crystal orientations. (b) Illustration of the three different positions that the hexagonal frame of SiC bilayers can assume in the lattice. (c) ABCABC...stacking sequence of cubic 3C-SiC (zinc blende structure).

SiC is a family of wide band gap (WBG) semiconductors, where organization of bilayers confers band gap energy exclusive to each polytype; 2.39 eV for 3C-, 3.265 eV for 4H- and 3.023 eV for 6H-SiC [1], [49]. These different properties have led to diverse applications across polytypes; 4H-SiC with the highest band gap is suited for power electronic devices, whilst 6H-SiC, due to a similar lattice constant to that of gallium nitride, is best suited in advancements in LEDs [122]

Alternatively, 3C-SiC heteroepitaxially deposited on silicon wafers has successfully been used in the generation of MEMS cantilevers [123]; with promising perspectives of further downscaling towards NEMS [60], [124]. SiC's explored bio- and hemo- compatibility make it an excellent candidate material for the for advancements in brain machine interfaces [4], [5]. The authors suggest [125] as an excellent review on advanced biomedical applications using SiC.

B. Synthesis of SiC

Recently, increased attention to SiC as a suitable material in numerous biomedical applications due to its WBG, chemical inertness and mechanical strength [126] has been realized as potential biotransducers in biosensors [4], [87], [125]. The main hexagonal polytypes of SiC can be grown as single-crystal ingots, from which SiC wafers can be obtained. Various methods are available to grow homoepitaxial single crystalline SiC films commonly through chemical vapour deposition (CVD) [118], [127] [128] [129] [130] and less commonly through liquid phase epitaxy (LPE) [131] and molecular beam epitaxy (MBE) [132]. The 3C-SiC polytype is the only one that can be grown hetero-epitaxially on silicon. This is particularly advantageous as fabrication on silicon is well established and widely available. Growth on silicon can typically be achieved through three steps: firstly removal of native oxides present through hydrogen surface etching, a carbonization step to bond C to the Si dangling bond to create the first SiC "buffer" layer, and finally using a Si:C ratio of 0.7 to grow a cubic single-crystal layer on the buffer layer [118]. Modification to these steps such as the inclusion of silane between carbonization and growth steps has led to the production of very high quality films [133]. Advancements in surface preparation have led to the development in numerous techniques including; oxidation [117], [134]; sublimation etching [120], [135]; photoelectrochemical etching [136]; chemomechanical polishing [137], [138]; plasma etching [139] and hydrogen etching [128], [140], [141].

Growth of heteroepitaxial 3C-SiC films on silicon with high quality is still to-date very challenging because of the large lattice and thermal mismatch between the films and the silicon substrate [3]. Poly-SiC on the other hand has been shown to be more versatile capable of growth on diverse substrates at lower temperatures (500-1200 °C) and has allowed for the generation of multiple growth protocols. Production of SiC nanoparticles is again different; an electrochemical method by Wu *et al.*, (2005) involved etching of polycrystalline 3C-SiC coupled to ultra -

sonication to yield an average particle size of 3.9 nm [142]. Similar electrochemical anodization etching coupled to mechanical grinding of nanoporous 6H-SiC also yielded 6H-SiC nanocrystals [143]. Lin *et al.*, (2008) used a low temperature, low pressure plasma reactor to synthesize amorphous SiC; subsequent annealing of the samples in argon at 800 °C yielded -SiC nanoparticles (<10 nm) and graphite [144]. The amorphous phase of SiC allows for adjustable Si and C stoichiometry. Alternatively, Leconte *et al.*, (2008) used inductively coupled plasma to control synthesis of -SiC nanopowders.Here, the SiC stoichiometry was controllable by the process pressure and the addition of methane to compensate the decarburization process [145]. Yang *et al.*, (2011) used laser ablation of Si in ethanol to produce 3C-SiC nanoparticles in a water suspension [146]. A study by Botsoa *et al.*, (2008) demonstrated the application of 3C-SiC quantum dots for living cell imaging. They synthesized 3C-SiC nanoparticles less than 10 nm in size through electrochemical anodization of polycrystalline wafers followed by successive grinding and finally centrifugation and were verified to have a non-toxic effect to the cell [147].

C. Immobilization of Biomolecules

Current molecular recognition systems require the immobilization of biomolecules through covalent attachment [134], [148], [149] with specific attention to structural order and composition to maintain biological activity to maintain sensitivity towards chemical stimuli [150], [151]. This happens through surface functionalization for controlling the surface chemistry of the substrate itself and is therefore specific to the selected semiconductor material. For example the surface functionalization of silicon dioxide (SiO_2) has been shown to be unsuitable in electrolytic solutions [152] as well as conferring high noise levels in field effect transmitters (FETs) [153]. A study by Spetz *et al.*, (2006) exploited the thermal resistance properties of SiC to fabricate FET gas base sensors that operate at ~1,000 °C [2]. A more recent paper by the same group built on this principle engineering a multifunctional sensor device with an integrated transistor and resonator, to measured NO metabolism in an individual's breath [154].

SiC has emerged as a promising biosensing material due also to the possibility of realizing a thin surface oxide, which is essential for the surface termination required to immobilize biomolecules. Passivation of the SiC surface can be realized through high quality monolayers with reactive sites generated either with a terminal hydrogen (H) or hydroxide (OH) surface [119], [128], [155], [156]. Theoretical studies by Preuss *et al.*, (2006) compared pyrrole-functionalized Si- and C-terminated SiC surfaces. They concluded that adsorption occurred through N-H dissociation and formation of N-Si bonding on the Si face however this was unstable on the C face most likely due to negative adsorption energy [157]. Similar *ab inito* computational chemical investigations chemical investigations by Cicero and Catellani also demonstrated a larger stability on the Si face of the SiC(001) surface compared to those on Si [158], [159]. A study by Schoell *et al.*, (2008) used self-assembled monolayers (SAMs) for functionalization of n-type 6H-SiC to the Si face. Covalent functionalization on patterned aminopropyldiethoxy-methylsilane (APDEMS) monolayers was later verified through immobilized fluorescently labelled proteins [155]. Williams *et al.*, (2012) also immobilized streptavidin via biotinylation on to APTES functionalized 4H-SiC (0001). To ensure immobilization; X-ray photoelectron spectroscopy (XPS), ellipsometry, contact angle and fluorescence microscopy was used to optimize APTES layer on the SiC surface. Notably, instead of performing a pirhana dip, hydroxylation occurred through oxygen plasma treatment with O_2: Ar (20:80) to grow a thin oxide layer [148]. Similar functionalization has been performed on 4H-SiC (0001) with mercaptopropyltrimethoxy-silane (MPTMS) and verified using XPS and water contact angle measurements [150]. Electrical contribution from APTES and MPTMS organic layers on 4H-SiC has been further evaluated and shown to exhibit a Schottky diode like I-V characteristic [134].

Numerous studies have been performed on the biosensing capabilities of the semi-insulating hexagonal polytypes (4H- & 6H-SiC) due to their low leakage, transparency and biochemical inertness [160]; however 3C-SiC has emerged as the polytype of choice. It is ideal for biomedical MEMs devices as it is less expensive and less polar when compared to 4H-, 6H-SiC [125], [161]. 3C-SiC has a well-defined surface for electron transfer due to a low gradient and in-plane stress from the closely packed cubic structure [162]. Coupled to high Young's modulus, 3C-SiC can be constructed into wide frequency resonators as cantilevers or bridge structures as detections methods for mass, gas and biomolecule identification.

D. SiC for Biosensing: Design Considerations

Extensive research has shown that SiC is a superior alternative to Si, due to its demonstrated biocompatibility [2], [4], [125], [134], [154], [163].

Moreover, as it can be deposited on silicon wafers, its micromachining into MEMS building blocks is straightforward. SiC Young's modulus are notably high (440 GPa), as well as breakdown field twice as large as Si (MVcm [136]; making SiC ideal to augment and even replace Si. In addition to this due to SiC chemical inertness [121], it is a suitable material in high temperatures, hostile environments and shows a high resistance to corrosion in body fluids [119], [128], [135], [155], [156], [164] [165] [166].SiC's high elastic modulus (424 GPa), high hardness (5.8 GPa) and low friction coefficient (0.17) make it ideal as an *in vivo* biosensor and smart implant [163], [167].Li *et al.*, (2005) used a nanoscale approach to characterize both the mechanical and tribological properties of SiC for orthopaedic applications [135]. Coupled to the electrical, mechanical and thermal properties of SiC reinforces its suitability as a biosensing substrate. These qualities alone, including the aforementioned WBG, increases the sensing capabilities of SiC as a semiconductor. Of the polytypes, 3C-SiC has demonstrated a major advantage in biosensing applications due to the reproducibility of quantitative surface modification and functionalization. This has been realized as a versatile format for immobilization of biomolecules with high reproducibility in bioanalytical applications.

Fabrication of silicon based micro- and nano-cantilevers are most commonly fabricated using a well -established top down approach comprising of four main techniques; film deposition; photolithography; etching and doping [167]. An important consideration is the design and shape of the cantilever sensors, and the way such sensors can be combined into a micro-array. For optical detection schemes both rectangular and T shaped cantilevers are commonly employed [168], whilst capacitive detection systems benefit from square pads leading to increased sensitivity [169]. On the other hand piezoresistive cantilevers are designed U shaped with a partial Wheatstone bridge circuit. A recent report by Kermany *et al.*, (2014) fabricated microresonators with mechanical quality factors (Q) over a million using highly stressed epitaxial SiC on silicon wafers. Such high Q-factors were achieved using perfect-clamped string structure using silicon surface micromachining processes [170]. This study also demonstrated the use of two photolithography steps and as such two different masks followed by xenon difluoride dry etching to release the microbeams *(see figure 4)*. There is extensive debate for the selection of the best material for fabrication of microcantilever structures. With increased downscaling trends towards nanostructures, SiC has been realized as an ideal choice due to its ratio of Young's modulus (E) to mass density (is significantly higher than other semiconducting materials (Si, Si_3N_4, SiO_2,

GaAs). Given the ratio $\sqrt{(E/\rho)}$ [124], [124], SiC demonstrates high fundamental resonance frequencies, combined with small force constants enabling high sensitivity to realize the potential of NEMS. Additionally SiC's excellent chemical stability allows for surface treatments to achieve higher Q factors [170], [171]. This is vital as the NEMS Q factor is governed by surface defects [124], [172]. As discussed, 3C-SiC is an ideal polytype for fabricating complex resonators [4], [126], [129], [130], [166], [170], [173]. Yang *et al.*, (2001) fabricated nanometer scale single 3C-SiC layers using dry etching to minimize damage caused by increased surface tension from wet etching. This study demonstrated that given the same geometry, nanometer-scale SiC resonators produce frequencies lower than GaAs and Si resonators [124]. Similarly increasing the residual stress of epitaxial 3C-SiC films enhances Q factors of SiC microstring resonators [170] outperforming the state-of-the-art based Si_3N_4 [174]. Si_3N_4 has long been the preferred resonator material, however, epitaxial 3C-SiC on silicon possesses additional advantages to this sensing platform. Unlike Si_3N_4, SiC is a semiconductor and thus can be doped [175], [176] and is piezoresistive [177]. Also note that sensors made of epitaxial 3C-SiC are expected to behave substantially better than poly-crystalline SiC sensors thanks to the absence of grain boundaries, and thus higher fracture resistance and resistance to corrosion. Finally, note that 3C-SiC is also a convenient alternative to Si_3N_4, as it can be used as solid source and template for direct transfer-free growth of graphene [178] [179] [180].

SECTION IV

GRAPHENE

Graphene is a one atom thick planar sheet of sp^2 bonded carbon atoms packed as a dense honeycomb crystal lattice with C-C bond length of 0.142 nm [181]. First isolated in 2004, through micromechanical cleavage of graphite [182]; graphene has revolutionized the nanotechnology platform as a next generation electronic sensing material. Monolayers and bilayers of graphene are zero-gap semiconductors; with only one electron charge carrier. With additional layers, several more charge carriers form, and as such the term graphene has been limited to 10 layers before being considered graphite. Graphene has generated considerable attention due to its, anomalous quantum Hall effect [183], [184], absence of localization

[185], chemical inertness [186], high thermal conductivity (5000 Wm (5000 $Wm^{-1}K^{-1}$) [187], high current density [188], [189], optical transmittance [190], [191], super hydrophobicity at a nanometer scale [192], high electron mobility at room temperature (250,000 cm^2/Vs) [190], [193] and extraordinary mechanical properties with Young's modulus above 1 TPa [182], [194].Additionally, reducing the dimensions of graphene to narrow ribbons the width of 1–2 nm, a distinct band gap can be attained, producing semiconductive graphene with unparalleled applications in transistors for nanoelectronics and high frequency applications [195], [196]. Graphene has been established as an outstanding conductor and plasmonic material. These properties allow for the augmentation of both MEMS and plasmonics label free methods as well as providing framework for further downscaling towards nanosensing devices.

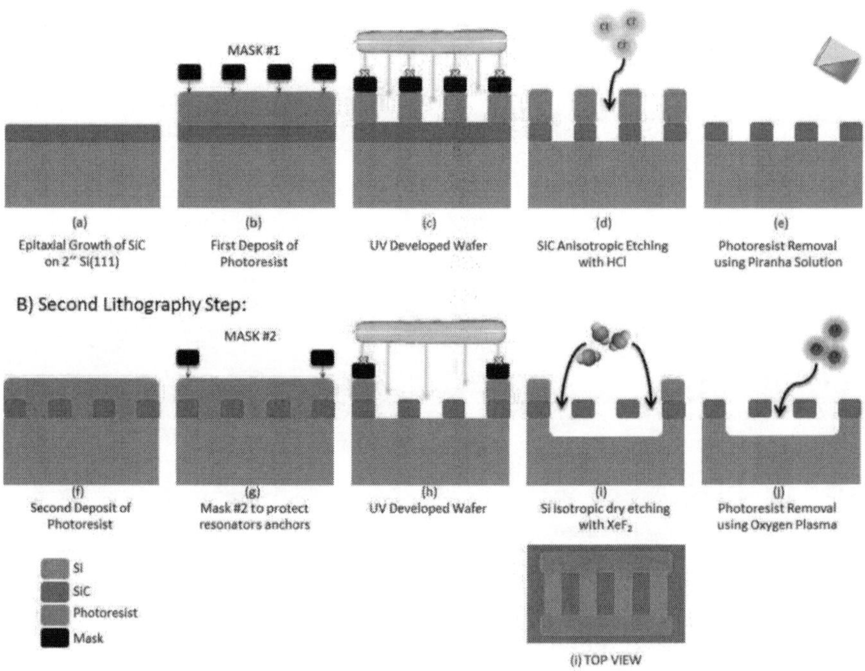

Figure 4. Two lithography steps using preferential etching techniques to fabricate perfect clamped epitaxial SiC on Si microstrings. Adapted from [168].

However, it is important to point out that at this stage, although the growth of 3C-SiC on silicon has seen substantial improvements, the interface of the SiC to the silicon substrate continues to be a problematic point. Additionally to being extremely defective because of the large mismatch stresses, it is also thermally unstable [F.Iacopi, unpublished data], which may prove an important limitation for some of the MEMS applications.

A. Biocompatibility of Graphene

The remarkable properties of graphene have stimulated increased investigations into its biocompatibility due to broad prospective applications in biomedical engineering and biotechnology. Graphene-based materials (GBMs) biocompatibility relies on their intrinsic physical-chemical properties, which alter due to the raw materials and fabrication methods, used [197], [198]. Liao *et al.*, (2011) attributed cytotoxicity to the particle size, quality, state, surface charge and oxygen threshold. It was concluded that GBMs are capable of inducing superoxide anion-independent oxidative stress on bacterial cells through oxidation of (γ) -L-glutamyl-L-cysteinyl-glycine [199]. In a study by Liu *et al.*, (2011) the toxicity effect of four GBMs (graphite (Gt), graphite oxide (GtO), graphene oxide (GO) and reduced graphene oxide (rGO)) was investigated against *E. coli*. Oxidative stress signals revealed GO had the highest antibacterial activities (>rGO>Gt>GtO) [200]. Such investigations have led to the antimicrobial applications of graphene.

Biofilm formation on conducting materials in the long-term use of bioimplants and biosensors is a significant problem [201]. Advancements in antimicrobial graphene film coating provides a prospective alternative to previously explored surface coatings such as antibiotics [202] and cationic peptides [203]. A report by Santos *et al.*, compared the *E. coli* growth rates on GO with graphene-poly-vinyl carbazole (PVK) nanocomposites. Results indicate greater than a 80% microbial inhibition from PVK treatment of GO due to an increase in solution dispersion and as such increased interaction with the bacteria [204]. A similar study by Carpio *et al.*, (2012) using PVK-GO nanocomposites revealed a strong antimicrobial effect to both gram negative and positive bacteria. Additional testing against fibroblast cells (NIH 3T3) revealed a significant neutral toxicity leading to numerous potential biomedical applications for the prevention of biofilm formation [205].

Other applications using manganese-ferrite ($MnFe_2O_4$) decorated GO nanocomposites have been shown to be ideal as contrast MRI agents and use in magnetic hyperthermia for cancer therapy. Further PEGylation of these composites demonstrated exceptional biocompatibility [206]. Recent studies by Lee *et al.*, explore graphene-based tissue engineering approaches using various types of stem cells to restore damaged or lost tissues. Here, graphene's aromatic scaffold and good biocompatibility provides support for stem cell growth and differentiation through non-covalent binding [192], [207].

B. Fabrication of Graphene

Several fabrication methods have been developed to produce graphene in large quantities the most common including micromechanical exfoliation of highly orientated pyrolytic graphite (HOPG) [182], CVD [208] [209] [210], Plasma enhanced CVD [211], [212] chemical reduction of graphite oxide (GO) [213], [214], CNTs unzipping [215] [216] [217] [218] and epitaxial growth on bulk SiC [178] [179] [180], [219], [220]. Mechanical exfoliation is simple in principle, relying on separation of layers of stacked graphene sheets from graphite to produce individual sheets. Novoselov *et al.*, (2004) subjected a 1 mm thick HOPG sheet to dry etching in oxygen plasma to make multiple deep plateaus. These were baked on a photoresist and subsequent peeled using scotch tape to remove layers from the graphite sheet. The resulting thin flakes were released from the photoresist in acetone and transferred to a Si substrate to yield mono to few-layer graphene sheets [182]. Though this first fabrication of graphene was indeed an exciting discovery, it is limited in its control over the number of layers produced and inability for large-scale production.

Accordingly, Stankovich *et al.*, (2007) proposed liquid phase exfoliation of graphite oxide due to its hydrophobicity for the large-scale production of graphene. This was achieved through exfoliation of GO nanosheets by ultrasonication in aqueous suspension and the subsequent reduction of these film in hydrazine hydrate at 100 °C for 24 hr [213]. However, Raman spectroscopic studies indicated that the invasive chemical treatment generated structural defects that disrupted the electronic structure of graphene [221]. It was later shown through XPS studies that even after chemical reduction or thermal annealing (up to 1000 °C), it is essentially impossible to regenerate the graphene structures [222].

To realize the promise for mass scale production of graphene, CVD approaches on metal surfaces provided a novel synthesis route for mono or few-layer graphene films. Somani *et al.*, (2006) first reported a thermal CVD (700-850 °C) technique to synthesize graphene from camphor on Ni foils. After cooling to ambient temperature yielded planar few layer graphene, high resolution transmission electron microscopy identified multiple folds and estimated to consist of 35 layers of graphene [210]. Whilst this method provides high quality graphene layers without complicated mechanical or chemical treatments, it still requires purification processes to eliminate the catalyst artifacts and transfer of graphene to another substrate. More recently, methods of transferring graphene using a micromanipulator and scanning electron microscope allow generation of GO nanoparticles that can be functionalized and used as biosensors. Xu *et al.*, (2013) demonstrated detection of trinitrotoluene (TNT) in 20 ppt using resonant microcantilever sensors functionalized with GO/Au nanoparticles (NP). The Au-NP serve as nanopillars to space GO sheets allowing molecules to access the nanopores. Hexafluoroisopropanol (HFIP) functionalized GO/Au-NPs hybrid material can be added to water to form a crude suspension, which can be loaded onto pre-prepared microcantilevers using a micromanipulator [229] *(see figure 5)*. However, transfer of graphene material requires precise manual handling techniques through the use of the micromanipulator and visualization with scanning electron microscopy (SEM) equipment, which can be expensive for smaller laboratories. Additionally, this method is effort- intensive, requiring serial functionalization of thousands of single cantilevers.

As stated graphene has a zero band gap and thus relies on introducing an energy band gap to be suitable for applications in semiconductors. This can be realized through either controlled oxidation of a few layers of graphene or fabrication of graphene nano ribbons (GNR). GNRs possess band gaps suitable for room temperature transistor operations with high carrier mobility due to narrow widths (<10 nm) and atomically smooth edges [217], [223], [224]. GNRs can be easily produced through e-beam lithography but this approach is limited by poor scale resolution (width of 20 nm) and large edge roughness [218], [225] [226] [227]. Accordingly, Jiao *et al.*, (2009) devised a novel approach through controlled unzipping of CNTs by argon plasma etching. This technique recognizes that CNTs are essentially GNRs rolled up into seamless tubes and therefore graphene growth can be controlled through the growth of CNTs. Multiwalled carbon nanotubes (MWCNTs) were embedded onto a poly (methyl methacrylate) (PMMA) layer on a Si substrate. This PMMA etch mask was purpose designed to leave a narrow strip of MWCNT sidewall exposed to facilitate faster etching. Subsequent argon plasma treatment lead to unzipped CNTs and smooth edged GNRs with uniform width of 10-20 nm corresponding to half the circumference of the starting MWCNT [217]. Various new unzipping technique have emerged facilitating large scale production GNRs with controlled structure; however, it is important to note that these materials have inferior electronic characteristics when compared to wide peeled sheets of graphene [181], [228].

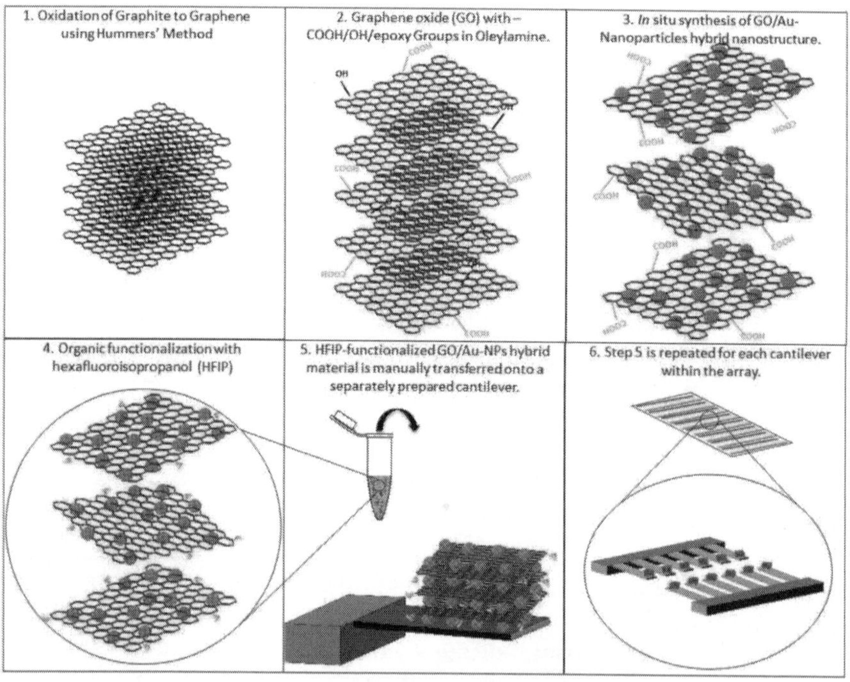

Figure 5. Gold (Au) – Nanoparticles (NP) *in situ* grown on bulk GO sheets to form GO/Au-NPs. Subsequent functionalization with hexafluoroisopropanol generates HFIP-functionalized GO/Au-NPs hybrid nanostructure suspensions that can be deposited onto pre-prepared cantilevers using a micromanipulator. Adapted from [229].

These fabrication methods to synthesize monolayer graphene and harness the extraordinary physical, optical, electronic and mechanical properties in actual micro- and nano-devices unfortunately still presents obstacles in the replication of well-defined structures when using transferred graphene [230], [231]. Precisely defined positions and dimensions is a requirement for the integration of graphene in MEMS and NEMS. Many methods reviewed require manipulation of exfoliated or grown graphene flakes and therefore are not considered suitable for augmenting such nanostructures. Encouragingly, a promising synthesis of homogenous, wafer size graphene for such technologies has been demonstrated by thermal decomposition of SiC [232] [233] [234].

C. Graphene From Silicon Carbide

Interestingly, silicon carbide is a suitable template as well as solid-source for the direct (transfer –free) growth of graphene on semiconductor wafers. As such, silicon carbide bulk substrates are an ideal platform for the synthesis of graphene for micro and nanodevices. Investigation into graphitization by annealing of SiC surfaces in ultrahigh vacuum began in 1975 [235]. Unfortunately the process was limited in scale and very expensive. More recently graphene coated structures from epitaxial SiC films on Si has been realized as promising templates to augment MEMS and NEMS applications. SiC is advantageous due to already established photolithography and etching patterning [178], coupled to solid carbon source processes for graphene growth [233], [236] [237] [238]. The use of established self-aligned approaches facilitates exacting dimensions within the nanometer range for well-defined graphitized structures. Rollings *et al.*, (2006) reported successful growth of single crystalline mono to few layer graphene films from thermal decomposition (1200 °C) of the (0001) Si face of the 6H-SiC wafer [236]. Growth of epitaxial graphene by *in vacuo* silicon sublimation from the (0001) and (000-1) faces of 4H- and 6H-SiC has also been shown [228]. Various reports have characterized similar graphene growth on SiC through high temperature annealing in vacuum methods [239] [240] [241], however this has been recently shown to yield graphene layers with small grains (30–200 nm) [242], [243].

To address this, a report by Emtsev *et al.*, (2009) introduced a different approach of *ex situ* graphitization of (0001) Si face of 6H SiC in an argon atmosphere of 1 bar. Raman spectroscopy and Hall measurements of these large-size monolayer graphene films demonstrated improved quality of film with high electronic motilities $(\mu = 2{,}000\ \mathrm{cm}^{2}\ \mathrm{V}^{-1}\ \mathrm{s}^{-1}$ at $T = 27\ \mathrm{K})$ [242]. Nickel-mediated catalytic graphitization at the SiC surface to obtain graphene has emerged as a possible route. The catalytic action of Ni is vital as it reacts with SiC at comparably low temperature forming $NiSi_4$ to mediate the release of carbon required for the synthesis of graphene [178], [234], [244], [245].Juang *et al.*, (2009) demonstrated low temperature (750 °C) synthesis of graphene on thin Ni films coated SiC substrates [237].

1) Graphitization of 3C-SiC

Graphitization of 3C-SiC on silicon is clearly a long sought-after goal, although it has historically been much more challenging. This is due to the lesser quality of the hetero-epitaxial SiC films on silicon, and the limitation of the graphene growth temperature to well below 1400 °C to remain within safe limits from the melting temperature of the silicon substrate. A suitable surface considered is 3C-SiC (111) as its top four layers are identical to those of 6H-SiC (0001), nevertheless challenges still arise from the high defects density of the hetero-epitaxial films due to the considerable lattice and thermal mismatch of Si and 3C- SiC [171], [246]. Many studies have demonstrated ultrathin graphene (monolayer, bilayer or multilayer) sheets from graphitization of 3C-SiC epilayers [247], [248] without needing to transfer the material to another insulating substrate before integration into MEMS and NEMS devices [249]. Finally graphitization of 3C-SiC (111) maintains a similar crystallographic structure that naturally accommodates the six-fold symmetry [250].

Investigations into reconstructions that lead to graphene growth have identified differences for the Si-terminated and C-terminated faces of SiC. On Si-terminated face scanning tunneling microscopy (STM) imagery has revealed a "zeroth layer graphene" as a $(6\sqrt{3} \times 6\sqrt{3})R30°$ reconstructed interface layer that serves as a precursor stage of graphitization [251]. As such monolayer graphene is the first layer growing on top of this buffer layer, due to desorption of Si atoms [179], [252], [253] *(see figure 6)*. Epitaxial graphene annealed at temperatures larger than 900 - 1200 °C under Si flux induce gradual development of (3×3), (6×6), $(\sqrt{3}\times\sqrt{3})$ R30° and finally the reconstructed interface $(6\sqrt{3} \times 6\sqrt{3})$ R30° layer. A count of carbon atoms demonstrates that three SiC bilayers provide the carbon source for one graphene layer. Therefore it has been assumed that reformation of $(6\sqrt{3} \times 6\sqrt{3})R30°$ reconstructed interface is required for each new graphene layer [254]. In this model the previous $(6\sqrt{3} \times 6\sqrt{3})$ R30° interface is released from its covalent bonding forming the next graphene layer. This interface ensures the same 30° rotation for all graphene layers grown on the SiC substrate.

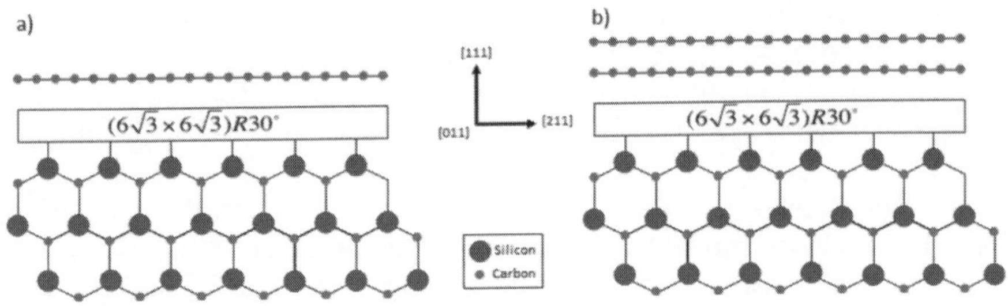

Figure 6. Cross sectional structural model along the [011] cubic SiC zone axis of (a) monolayer and (b) bilayer epitaxial graphene on 3C-SiC(111). The graphene is growing on top of the $(6\sqrt{3} \times 6\sqrt{3})$ R30° reconstructed interface layer.

Alternatively, on the C-terminated face the $(6\sqrt{3} \times 6\sqrt{3})$ R30° reconstruction is not observed, and the atomic transformation of graphene not as easily clarified. Studies by Gupta *et al.*, using XPS, STM and Raman spectroscopy reveal interesting details on the transformation from 3C SiC (111)/Si (111) to graphene at different temperatures (1125 to 1375 °C). STM imaging revealed continuous hexagonal structure mono to few-layer graphene with hole-to-hole dimensions of 0.246 nm. Images also demonstrate transition from 3C SiC (111)/Si (111) to graphene at 1250 °C occurred in two subsequent steps: transition from SiC surface to followed by monolayer graphene. Wrinkled areas were attributed to defects in the underlying SiC/Si (111) and remain an issue of lattice mismatch. Raman spectroscopy indicates that G band shifts towards lower wavelengths, and as such an increase in graphene layer thickness is a strong function of increases in annealing temperature [179], [180].Finally this investigation noted that 3C-SiC does not depict second order Raman features between 1450 and 1750 cm as in other polytypes of SiC [255].

Synthesis of graphene on epitaxial SiC on Si (briefly discussed earlier) by Cunning *et al.*, (2014) demonstrate uniform and high- quality graphene synthesis using a Ni-Cu alloy catalyst. A pre-patterned 3C-SiC layer (250 nm) on a Si (111) wafer was sputtered with nickel and copper. Subsequent site selective graphitization at 1000 °C through; (a)

Kirkendall diffusion of the nickel and copper; (b) silicide formation to generate a carbon source for graphitization; (c) removal of intermixed metal to expose the graphene on the SiC through sonication in a Freckle's etch solution [178] *(see figure 7)*. This self-aligned nanocoating of few layer graphene onto suspended SiC microstructures is ideal for replacing current conducting layer in MEMS, as well as providing alternate surface functionalization routes. This novel synthesis approach is also advantageous as it can be performed at temperatures compatible with conventional semiconductor processing, allowing complex device integration and generation of thousands of MEMS/NEMS by pre-patterning of the SiC. A follow up study by Iacopi *et al.*, (2015) introduced novel synthesis of high quality and highly uniform few layer graphene from epitaxial 3C-SiC films on Si (100) and Si (111) using the described [178] Ni/Cu catalytic alloy approach. Transmission electron microscopy in this study [232] yielded <0.9 nm thin bilayer graphene with high-quality and high adhesion to the substrate [256]. Such advances open enormous new opportunities for harnessing the properties of graphene for MEMS and NEMS devices fabricated at the wafer -level.

D. Functionalization of Graphene

Graphene can be prepared readily through chemical, thermal or photo-chemical reduction of GO, and this is the type of graphene most heavily investigated so far for functionalisation. However, subsequent graphene formation can lead to precipitation of graphite particles and sheet aggregation due to lack of a stabilizer [257]. As such covalent or non-covalent surface functionalization is performed before reducing the graphene sheets [258]. Fortunately this process enables the immobilization of biomolecules through surface functionalization of highly oxygenated GO sheets.Covalent functionalization of this framework occurs through hybridization of one or more sp^2 carbon atoms into the sp^3 configuration [259] through either nucleophilic substitution [260], [261] electrophilic addition [262], [263], condensation reactions [264], [265] or addition reactions [266], [267].

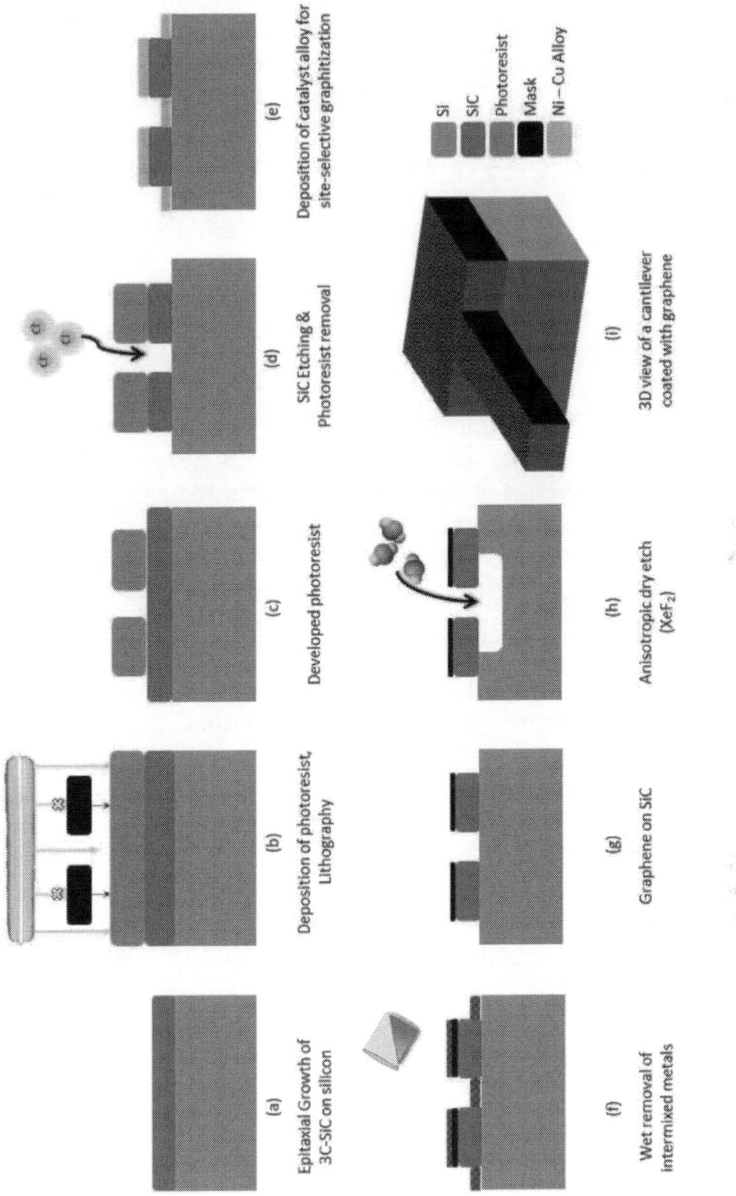

Figure 7. Wafer-level fabrication of graphitized silicon carbide microbeams on a silicon substrate. The SiC can be patterned to generate thousands of cantilevers and thus removes the need for manipulation of single flakes. The few-layer graphene is grown selectively on the SiC structures via site-selected graphitization. Adapted from [176].

Nucleophilic substitution reactions are commonly employed for the functionalization of GO as it occurs easily at room temperature and in an aqueous medium. Organic modifiers containing the amine functionality can undergo nucleophilic substitution reactions that target the epoxy groups of GO. Many aliphatic and aromatic amines, amino acids, amine-terminated biomolecules and small molecular weight polymers have been successfully prepared on functionalized graphene [257]. For example, Kuila*et al* (2010) used dodecyl amine (DA) and octadecyl amine (ODA) for the surface treatment of graphene, [268] the results being similar to those of Bourlinos *et al.*, (2003) who demonstrated that amine intercalated GO derivatives depend on amine chain length [261]. Chemical functionalization of graphene sheets can also be carried out with APTES [260]. Similar APTES chemistry has previously been described on SiC platforms and as such could provide a suitable framework to integrate graphene into MEMS systems. A study by Shan *et al.*, (2009) generated biocompatible graphene functionalized with poly-L-lysine (PLL) to immobilize biomolecules. The resulting biosensor was conjugated to horseradish peroxide (HRP) to create graphene-PLL/HRP nanocomposites capable of sensing H_2O_2 [269]. Graphene functionalization through condensation reactions has been shown to occur through loss of entropy. Recent investigations have demonstrated that condensation occurs with isocyanate, diisocyanate, and amine compounds through the formation of amides and carbonate ester linkages. Stankovich *et al.*, (2006) functionalized graphene using isocyanate compounds revealed reduced hydrophilic properties, which through further exfoliation yielded derivatized, GO nanoplatelets [264]. Diisocyantes have also been used for graphene functionalization by activating the carboxyl functionality using thionyl chloride ($SOCl_2$). This nanoporous material displayed promise as a hydrogen adsorbent material [270]. Finally, Liu *et al.*, (2008) first reported GO functionalization using poly(ethylene glycol) ($PEG-NH_2$) through a carbodiimide catalyzed amide formation.Various insoluble aromatic drug molecules such as camptothecin analogues and Iressa (Gefitinib) were successfully loaded onto nanoscale graphene oxide-PEG via simple adsorption [265], [271]. Further studies to optimize functionalization of graphene for biomolecule immobilization highlights a promising pathway for increased sensitivity and NEMS downscaling.

SECTION V

CONCLUSION

Label free, analyte specific, accurate, and highly sensitive biochemical detection systems will offer superior capabilities over existing systems in terms of high-throughput, efficient drug screening and early pathogen detection. Research in this area, underpinned by novel functional (nano)materials, is now reaching maturity and offers perspectives of concrete advances. Silicon carbide, potentially combined with transfer – free graphene, is one prominent example of material system that could provide enormous advances, either through plasmonics or micro-electro-mechanical systems. Additionally, the possibility for scaling down the sizes of such sensors towards the nanoscale, together with ensuring the bio- and hemo-compatibility of the materials involved, could potentially enable minimally invasive endoscopic detection, with enormous benefits for early detection of life-threatening diseases such as cancer.

ACKNOWLEDGMENT

O.C thanks Atieh R. Kermany and Neeraj Mishra for their invaluable input and guidance throughout the preparation of this manuscript.

FOOTNOTES

This work was supported in part by the Air Force Office of Scientific Research, Office of Naval Research Global, and Army Research, Development and Engineering Command through the AFOSR/Asian Office of Aerospace Research and Development under Grant 15IOA053 and the Australian National Fabrication Facility, Queensland Node. The work of F. Iacopi was supported by the Australian Research Council under Grant FT120100445.

REFERENCES

1. P. M. Sarro "Silicon carbide as a new MEMS technology" Sens. Actuators A, Phys., vol. 82, no. 3, pp. 210-218, *May, 2000*

2. L. Spetz et al "New materials for chemical and biosensors" Mater. Manuf. Process., vol. 21, no. 3, pp. 253-256, *May, 2006*

3. F. Iacopi, M. Van Hove, M. Charles and K. Endo "Power electronics with wide bandgap materials: Toward greener, more efficient technologies" MRS Bull., vol. 40, no. 5, pp. 390-395, *2015*

4. L. Frewin, C. Locke, S. E. Saddow and E. J. Weeber "Single-crystal cubic silicon carbide: An *in vivo* biocompatible semiconductor for brain machine interface devices" Proc. Annu. Int. Conf. IEEE Eng. Med. Biol. Soc., pp. 2957-2960, *Aug./Sep. 2011*

5. L. Frewin et al "Atomic force microscopy analysis of central nervous system cell morphology on silicon carbide and diamond substrates" J. Molecular Recognit., vol. 22, no. 5, pp. 380-388, *2009*

6. S. E. Saddow, C. L. Coletti, C. Frewin, N. C. Schettini, A. Oliveros and M. Jaroszeski "Single-crystal silicon carbide: A biocompatible and hemocompatible semiconductor for advanced biomedical applications" Proc. MRS Symp., vol. 679, pp. 824-830, *Feb., 2011*

7. Coletti, M. J. Jaroszeski, A. Pallaoro, A. M. Hoff, S. Iannotta and S. E. Saddow "Biocompatibility and wettability of crystalline SiC and Si surfaces" Proc. 29th Annu. Int. Conf. IEEE Eng. Med. Biol. Soc., pp. 5849-5852, *Aug. 2007*

8. C. L. Frewin et al "The development of silicon carbide based electrode devices for central nervous system biomedical implants" Proc. MRS Symp., vol. 1236, p. 1236-SS01-02, *Jan. 2011*

9. S. Saddow Silicon Carbide Biotechnology: A Biocompatible Semiconductor for Advanced Biomedical Devices and Applications *2011*, Elsevier, Amsterdam, The Netherlands

10. S. Dai et al "Graphene on hexagonal boron nitride as a tunable hyperbolic metamaterial" Nature Nanotechnol., vol. 10, no. 8, pp. 682-686, *Aug., 2015*

11. J. T. Kemp, R. W. Davis, R. L. White, S. X. Wang and C. D. Webb "A novel method for STR-based DNA profiling using microarrays" J. Forensic Sci., vol. 50, no. 5, pp. 1109-1113, *Sep., 2005*

12. C. P. Paweletz et al "Reverse phase protein microarrays which capture disease progression show activation of pro-survival pathways at the cancer invasion front" Oncogene, vol. 20, no. 16, pp. 1981-1989, *Apr., 2001*

13. S. Nishizuka et al "Proteomic profiling of the NCI-60 cancer cell lines using new high-density reverse-phase lysate microarrays" Proc. Nat. Acad. Sci. USA, vol. 100, no. 24, pp. 14229-14234, *Nov., 2003*

14. C. J. Day, G. Tram, L. E. Hartley-Tassell, J. Tiralongo and V. Korolik "Assessment of glycan interactions of clinical and avian isolates of Campylobacter jejuni" BMC Microbiol., vol. 13, no. 1, p. 1, *Oct., 2013*

15. P.-H. Liang, C.-Y. Wu, W. A. Greenberg and C.-H. Wong "Glycan arrays: Biological and medical applications" Curr. Opin. Chem. Biol., vol. 12, no. 1, pp. 86-92, *Dec., 2007*

16. N. X. Arndt et al "Differential carbohydrate binding and cell surface glycosylation of human cancer cell lines" J. Cell. Biochem., vol. 112, no. 9, pp. 2230-2240, *Aug., 2011*

17. S. C. Tao et al "Lectin microarrays identify cell-specific and functionally significant cell surface glycan markers" Glycobiology, vol. 18, no. 10, pp. 761-769, *Sep., 2008*

18. R. Akbani et al "Realizing the promise of reverse phase protein arrays for clinical, translational, and basic research: A workshop report: The RPPA (reverse phase protein array) society" Molecular Cellular Proteomics, vol. 13, no. 7, pp. 1625-1643, *Jul., 2014*

19. Hall, J. Ptacek and M. Snyder "Protein microarray technology" Mech. Ageing Develop., vol. 128, no. 1, pp. 161-167, *Jan., 2007*

20. N. Ramachandran, D. N. Larson, P. R. H. Stark, E. Hainsworth and J. LaBaer "Emerging tools for real-time label-free detection of interactions on functional protein microarrays" FEBS J., vol. 272, no. 21, pp. 5412-5425, *Nov., 2005*

21. K. A. Willets and R. P. Van Duyne "Localized surface plasmon resonance spectroscopy and sensing" Annu. Rev. Phys. Chem., vol. 58, pp. 267-297, *May, 2007*

22. Otto "Excitation of nonradiative surface plasma waves in silver by the method of frustrated total reflection" Zeitschrift Phys., vol. 216, pp. 398-410, *Aug., 1968*

23. J. Homola "Surface plasmon resonance sensors for detection of chemical and biological species" Chem. Rev., vol. 108, no. 2, pp. 462-493, *Feb., 2008*

24. J. Homola, S. S. Yee and G. Gauglitz "Surface plasmon resonance sensors: Review" Sens. Actuators B, Chem., vol. 54, no. 1, pp. 3-15, *Dec., 199*

25. Choi and Y. Choi "Plasmonic nanosensors: Review and prospect" IEEE J. Sel. Topics Quantum Electron., vol. 18, no. 3, pp. 1110-1121, *May/Jun., 2012*

26. J. G. Gordon, II and S. Ernst "Surface plasmons as a probe of the electrochemical interface" Surf. Sci., vol. 101, no. 1–3, pp. 499-506, *Jan., 1980*

27. S. M. Tabakman et al "Plasmonic substrates for multiplexed protein microarrays with femtomolar sensitivity and broad dynamic range" Nature Commun., vol. 2, p. 466, *Sep., 2011*

28. N. Anker, W. P. Hall, O. Lyandres, N. C. Shah, J. Zhao and R. P. Van Duyne

29. "Biosensing with plasmonic nanosensors" Nature Mater., vol. 7, no. 6, pp. 442-453, *May, 2008*

30. A.G. Brolo, R. Gordon, B. Leathem and K. L. Kavanagh "Surface plasmon sensor based on the enhanced light transmission through arrays of nanoholes in gold films" Langmuir, vol. 20, no. 12, pp. 4813-4815, *Jun., 2004*

31. T. Kang, J. Moon, S. Oh, S. Hong, S. Chah and J. Yi "Direct observation of a cooperative mechanism in the adsorption of heavy metal ions to thiolated surface by in-situ surface plasmon resonance measurements" Chem. Commun., vol. 18, no. 18, pp. 2360-2362, *May, 2005*

32. H. Im, A. Lesuffleur, N. C. Lindquist and S.-H. Oh "Plasmonic nanoholes in a multichannel microarray format for parallel kinetic assays and differential sensing" Anal. Chem., vol. 81, no. 8, pp. 2854-2859, *2009*

33. R. Yonzon, E. Jeoung, S. Zou, G. C. Schatz, M. Mrksich and R. P. Van Duyne "A comparative analysis of localized and propagating surface plasmon resonance sensors: The binding of concanavalin a to a monosaccharide functionalized self-assembled monolayer" J. Amer. Chem. Soc., vol. 126, no. 39, pp. 12669-12676, *Oct., 2004*

34. T. Kang, S. Oh, S. Hong, J. Moon and J. Yi "Mesoporous silica thin films as a spatially extended probe of interfacial electric fields for amplified signal transduction in surface plasmon resonance spectroscopy" Chem. Commun., vol. 28, no. 28, pp. 2998-3000, *Jul., 2006*

35. T. R. Jensen, M. D. Malinsky, C. L. Haynes and R. P. Van Duyne

36. "Nanosphere lithography: Tunable localized surface plasmon resonance spectra of silver nanoparticles" J. Phys. Chem. B, vol. 104, no. 45, pp. 10549-10556, *Nov., 2000*

37. T. Kang et al "Reversible pH-driven conformational switching of tethered superoxide dismutase with gold nanoparticle enhanced surface plasmon resonance spectroscopy" J. Amer. Chem. Soc., vol. 128, no. 39, pp. 12870-12878, *Oct., 2006*

38. P. L. Stiles, J. A. Dieringer, N. C. Shah and R. P. Van Duyne "Surface-enhanced Raman spectroscopy" Annu. Rev. Anal. Chem., vol. 1, no. 1, pp. 601-626, *Jul., 2008*

39. J. Haes, S. Zou, G. C. Schatz and R. P. Van Duyne "A nanoscale optical biosensor: The long range distance dependence of the localized surface plasmon resonance of noble metal nanoparticles" J. Phys. Chem. B, vol. 108, no. 1, pp. 109-116, *2004*

40. McFarland, M. A. Young, J. A. Dieringer and R. P. Van Duyne "Wavelength-scanned surface-enhanced Raman excitation spectroscopy" J. Phys. Chem. B, vol. 109, no. 22, pp. 11279-11285, *Jun., 2005*

41. R. Karlsson "SPR for molecular interaction analysis: A review of emerging application areas" J. Molecular Recognit., vol. 17, no. 3, pp. 151-161, *May/Jun., 2004*

42. V. Kanda, J. K. Kariuki, D. J. Harrison and M. T. McDermott "Label-free reading of microarray-based immunoassays with surface plasmon resonance imaging" Anal. Chem., vol. 76, no. 24, pp. 7257-7262, *Dec., 2004*

43. J. S. Yuk et al "Analysis of protein interactions on protein arrays by a novel spectral surface plasmon resonance imaging" Biosensors Bioelectron., vol. 21, no. 8, pp. 1521-1528, *2006*

44. M. Medina-Sánchez, S. Miserere and A. Merkoçi "Nanomaterials and lab-on-a-chip technologies" Lab Chip, vol. 12, no. 11, pp. 1932-1943, *May, 2012*

45. A.V. Whitney et al "Localized surface plasmon resonance nanosensor: A high-resolution distance-dependence study using atomic layer deposition" J. Phys. Chem. B, vol. 109, no. 43, pp. 20522-20528, *Nov., 2005*

46. S. Kim, J.-M. Jung, D.-G. Choi, H.-T. Jung and S.-M. Yang "Patterned arrays of Au rings for localized surface plasmon resonance" Langmuir, vol. 22, no. 17, pp. 7109-7112, *2006*

47. M. Larsson, J. Alegret, M. Käll and D. S. Sutherland "Sensing characteristics of NIR localized surface plasmon resonances in gold nanorings for application as ultrasensitive biosensors" Nano Lett., vol. 7, no. 5, pp. 1256-1263, *Apr., 2007*

48. L. A. Dick, A. J. Haes and R. P. Van Duyne "Distance and orientation dependence of heterogeneous electron transfer: A surface-enhanced resonance Raman scattering study of cytochrome c bound to carboxylic acid terminated alkanethiols adsorbed on silver electrodes" J. Phys. Chem. B, vol. 104, no. 49, pp. 11752-11762, *2000*

49. Ruach-Nir, T. A. Bendikov, I. Doron-Mor, Z. Barkay, A. Vaskevich, and I. Rubinstein, "Silica-stabilized gold island films for transmission localized surface plasmon sensing," J. Amer. Chem. Soc., vol. 129, no. 1, pp. 84–92, Jan. 2007.

50. P. S. Waggoner and H. G. Craighead, "Micro- and nanomechanical sensors for environmental, chemical, and biological detection," Lab Chip, vol. 7, no. 10, p. 1238, 2007.

51. W. J. Choyke and G. Pensl, "Physical properties of SiC," MRS Bull., vol. 22, no. 3, pp. 25–29, Mar. 1997.

52. G. Wu, R. H. Datar, K. M. Hansen, T. Thundat, R. J. Cote, and A. Majumdar, "Bioassay of prostate-specific antigen (PSA) using microcantilevers," Nature Biotechnol., vol. 19, no. 9, pp. 856–860, Aug. 2001.

53. P. J. Hesketh, G. Barna, and H. G. Hughes, Microstructures and Microfabricated Systems III. Pennington, NJ, USA: ECS, 1997.

54. J. Gamby, A. Rudolf, M. Abid, H. H. Girault, C. Deslouis, and B. Tribollet, "Polycarbonate microchannel network with carpet of gold nanowires as SERS-active device," Lab Chip, vol. 9, no. 12, pp. 1806–1808, Jun. 2009.

55. P. R. West, S. Ishii, G. V. Naik, N. Emani, V. M. Shalaev, and A. Boltasseva, "Searching for better plasmonic materials," Laser Photon. Rev., vol. 4, no. 6, pp. 795–808, Nov. 2009.

56. M. G. Blaber, M. D. Arnold, and M. J. Ford, "Designing materials for plasmonic systems: The alkaliŰnoble intermetallics," J. Phys., Condens. Matter, vol. 22, no. 9, p. 095501, 2010.

57. Boltasseva and H. A. Atwater, "Low-loss plasmonic metamaterials," Mater. Sci., vol. 331, no. 6015, pp. 290–291, Jan. 2011. [56] R.-M. Ma, R. F. Oulton, V. J. Sorger, and X. Zhang, "Plasmon lasers: Coherent light source at molecular scales," Laser Photon. Rev., vol. 7, no. 1, pp. 1–21, Feb. 2012.

58. H. Wei and H. Xu, "Plasmonics in composite nanostructures," Biochem. Pharmacol., vol. 17, no. 8, pp. 372–380, Oct. 2014.

59. O. J. Glembocki et al., "Beyond electron based metamaterials: Lowloss surface phonon polariton-based nano-antenna arrays using silicon carbide," in Proc. ECS Meeting Abstracts, pp. 1548–1548, 2014.

60. K. Brueckner et al., "Micro- and nano-electromechanical resonators based on SiC and group III-nitrides for sensor applications," Phys. Status Solidi A, vol. 208, no. 2, pp. 357–376, Jul. 2010.

61. M. L. Roukes, "Nanoelectromechanical systems," in Tech. Digest 2000 Solid-State Sen. Actuat. Workshop, Hilton Head Island, SC, pp. 1–10, 2000.

62. P.-S. Lee et al., "Microcantilevers with nanochannels," Adv. Mater., vol. 20, no. 9, pp. 1732–1737, 2008.

63. P. I. Oden, G. Y. Chen, R. A. Steele, R. J. Warmack, and T. Thundat, "Viscous drag measurements utilizing microfabricated cantilevers," Appl. Phys. Lett., vol. 68, no. 26, pp. 3814–3816, 1996.

64. G. Meyer and N. M. Amer, "Novel optical approach to atomic force microscopy," Appl. Phys. Lett., vol. 53, no. 12, pp. 1045–1047, 1988.

65. R. Datar et al., "Cantilever sensors: Nanomechanical tools for diagnostics," MRS Bull., vol. 34, no. 6, pp. 449–454, Jun. 2009.

66. H. Ibach, "Adsorbate-induced surface stress," J. Vac. Sci. Technol. A, vol. 12, no. 4, p. 2240, 1994.

67. T. Thundat, L. Pinnaduwage, and R. Lareau, "Explosive vapour detection using micromechanical sensors," Abstr. Pap. Am. Chem. S., vol. 159, pp. 249–266, 2004.

68. G. G. Stoney, "The tension of metallic films deposited by electrolysis," Proc. Roy. Soc. London A, vol. 82, no. 553, pp. 172–175, May 1909.

69. K. S. Hwang, J. H. Lee, J. Park, D. S. Yoon, J. H. Park, and T. S. Kim, "In-situ quantitative analysis of a prostate-specific antigen (PSA) using a nanomechanical PZT cantilever," Lab Chip, vol. 4, no. 6, pp. 547–552, Nov. 2004.

70. H. P. Lang, M. Hegner, and C. Gerber, "Nanomechanics—The link to biology and chemistry," CHIMIA Int. J. Chem., vol. 56, no. 10, pp. 515–519, Oct. 2002.

71. K. E. Petersen and C. R. Guarnieri, "Young's modulus measurements of thin films using micromechanics," J. Appl. Phys., vol. 50, no. 11, p. 6761, 1979.

72. K. M. Goeders, J. S. Colton, and L. A. Bottomley, "Microcantilevers: Sensing chemical interactions via mechanical motion," Chem. Rev., vol. 108, no. 2, pp. 522–542, Feb. 2008.

73. F. R. Blom, S. Bouwstra, M. Elwenspoek, and J. H. J. Fluitman, "Dependence of the quality factor of micromachined silicon beam resonators on pressure and geometry," J. Vac. Sci. Technol. B, vol. 10, no. 1, pp. 19–26, Jan. 1992.

74. J. Tamayo, "Study of the noise of micromechanical oscillators under quality factor enhancement via driving force control," J. Appl. Phys., vol. 97, no. 4, p. 044903, 2005.

75. J. Fritz et al., "Translating biomolecular recognition into nanomechanics," Science, vol. 288, no. 5464, pp. 316–318, Apr. 2000.

76. Ilic, Y. Yang, K. Aubin, R. Reichenbach, S. Krylov, and H. G. Craighead, "Enumeration of DNA molecules bound to a nanomechanical oscillator," Nano Lett., vol. 5, no. 5, pp. 925–929, Apr. 2005.

77. Boisen, J. Thaysen, H. Jensenius, and O. Hansen, "Environmental sensors based on micromachined cantilevers with integrated read-out," Ultramicroscopy, vol. 82, nos. 1–4, pp. 11–16, Dec. 1999.

78. R. Mukhopadhyay, M. Lorentzen, J. Kjems, and F. Besenbacher, "Nanomechanical sensing of DNA sequences using piezoresistive cantilevers," Langmuir, vol. 21, no. 18, pp. 8400–8408, Aug. 2005.

79. K. W. Wee et al., "Novel electrical detection of label-free disease marker proteins using piezoresistive self-sensing micro-cantilevers," Biosensors Bioelectron., vol. 20, no. 10, pp. 1932–1938, Apr. 2005.

80. G. Binnig, C. F. Quate, and C. Gerber, "Atomic force microscope," Phys. Rev. Lett., vol. 56, no. 9, pp. 930–933, 1986.

81. N. V. Lavrik, M. J. Sepaniak, and P. G. Datskos, "Cantilever transducers as a platform for chemical and biological sensors," Rev. Sci. Instrum., vol. 75, no. 7, pp. 2229–2253, Jul. 2004.

82. L. Britton, Jr., et al., "Multiple-input microcantilever sensors," Ultramicroscopy, vol. 82, nos. 1–4, pp. 17–21, Feb. 2000.

83. M. Zougagh and A. Ríos, "Micro-electromechanical sensors in the analytical field," Analyst, vol. 134, no. 7, pp. 1274–1290, 2009.

84. K. M. Hansen et al., "Cantilever-based optical deflection assay for discrimination of dna single-nucleotide mismatches," Anal. Chem., vol. 73, no. 7, pp. 1567–1571, Apr. 2001.

85. G. Stemme, "Resonant silicon sensors," J. Micromech. Microeng., vol. 1, no. 2, pp. 113–125, Jun. 1991.

86. P. A. Rasmussen, J. Thaysen, O. Hansen, S. C. Eriksen, and A. Boisen, "Optimised cantilever biosensor with piezoresistive read-out," Ultramicroscopy, vol. 97, nos. 1–4, pp. 371–376, Oct. 2003.

87. G. A. Campbell and R. Mutharasan, "Piezoelectric-excited millimetersized cantilever (PEMC) sensors detect Bacillus anthracis at 300 spores/mL," Biosensors Bioelectron., vol. 21, no. 9, pp. 1684–1692, Mar. 2006.

88. G. Kotzar et al., "Evaluation of MEMS materials of construction for implantable medical devices," Biomaterials, vol. 23, no. 13, pp. 2737–2750, Jun. 2002.

89. S. S. Lee and R. M. White, "Self-excited piezoelectric cantilever oscillators," Sens. Actuators A, Phys., vol. 52, nos. 1–3, pp. 41–45, Dec. 1995.

90. J. H. Lee, K. H. Yoon, K. S. Hwang, J. Park, S. Ahn, and T. S. Kim, "Label free novel electrical detection using micromachined PZT monolithic thin film cantilever for the detection of C-reactive protein," Biosensors Bioelectron., vol. 20, no. 2, pp. 269–275, Sep. 2004.

91. R. S. Wagner and W. C. Ellis, Vapor-Liquid-Solid Mechanism of Crystal Growth and Its Application to Silicon. New York, NY, USA: AIME, 1965.

92. L. Latu-Romain and M. Ollivier, "Silicon carbide based one-dimensional nanostructure growth: Towards electronics and biology perspectives," J. Phys. D, Appl. Phys., vol. 47, no. 20, p. 203001, May 2014.

93. S. A. Fortuna and X. Li, "TOPICAL REVIEW: Metal-catalyzed semiconductor nanowires: A review on the control of growth directions," Semicond. Sci. Technol., vol. 25, no. 2, p. 024005, Feb. 2010.

94. R. Rurali, "Colloquium: Structural, electronic, and transport properties of silicon nanowires," Rev. Mod. Phys., vol. 82, no. 1, pp. 427–449, Jan. 2010.

95. L. J. Lauhon, M. S. Gudiksen, D. Wang, and C. M. Lieber, "Epitaxial core–shell and core–multishell nanowire heterostructures," Nature, vol. 420, no. 6911, pp. 57–61, Nov. 2002.

96. Y. Cui, Q. Wei, H. Park, and C. M. Lieber, "Nanowire nanosensors for highly sensitive and selective detection of biological and chemical species," Science, vol. 293, no. 5533, pp. 1289–1292, Aug. 2001.

97. F. Patolsky, G. Zheng, O. Hayden, M. Lakadamyali, X. Zhuang, and C. M. Lieber, "Electrical detection of single viruses," Proc. Nat. Acad. Sci. USA, vol. 101, no. 39, pp. 14017–14022, Sep. 2004.

98. X. Song, B. Xia, S. R. Stowell, Y. Lasanajak, D. F. Smith, and R. D. Cummings, "Novel fluorescent glycan microarray strategy reveals ligands for galectins," Chem. Biol., vol. 16, no. 1, pp. 36–47, Jan. 2009.

99. M. Safi et al., "Interactions between magnetic nanowires and living cells: Uptake, toxicity, and degradation," ACS Nano, vol. 5, no. 7, pp. 5354–5364, Jul. 2011.

100. S. Iijima, "Helical microtubules of graphitic carbon," Nature, vol. 354, pp. 56–58, Nov. 1991.

101. J. Okuno, K. Maehashi, K. Kerman, Y. Takamura, K. Matsumoto, and E. Tamiya, "Label-free immunosensor for prostate-specific antigen based on single-walled carbon nanotube array-modified microelectrodes," Biosensors Bioelectron., vol. 22, nos. 9–10, pp. 2377–2381, Apr. 2007.

102. J.-C. Charlier, X. Blase, and S. Roche, "Electronic and transport properties of nanotubes," Rev. Mod. Phys., vol. 79, no. 2, pp. 677–732, May 2007.

103. S.-P. Huang et al., "First-principles study: Size-dependent optical properties for semiconducting silicon carbide nanotubes," Opt. Exp., vol. 15, no. 17, pp. 10947–10957, Aug. 2007.

104. Pham-Huu, N. Keller, G. Ehret, and M. J. Ledoux, "The first preparation of silicon carbide nanotubes by shape memory synthesis and their catalytic potential," J. Catalysis, vol. 200, no. 2, pp. 400–410, Jun. 2001.

105. J.-X. Zhao and Y.-H. Ding, "Silicon carbide nanotubes functionalized by transition metal atoms: A density-functional study," J. Phys. Chem. C, vol. 112, no. 7, pp. 2558–2564, Feb. 2008.

106. X. Wang and K. M. Liew, "Density functional study of fluorinated singlewalled silicon carbide nanotubes," J. Phys. Chem. C, vol. 116, no. 2, pp. 1702–1708, Jan. 2012.

107. Y. Arntz et al., "Label-free protein assay based on a nanomechanical cantilever array," Nanotechnology, vol. 14, no. 1, pp. 86–90, Jan. 2003.

108. K. A. Stevenson, A. Mehta, P. Sachenko, K. M. Hansen, and T. Thundat, "Nanomechanical effect of enzymatic manipulation of DNA on microcantilever surfaces," Langmuir, vol. 18, no. 23, pp. 8732–8736, Nov. 2002.

109. Ilic, D. Czaplewski, H. G. Craighead, P. Neuzil, C. Campagnolo, and C. Batt, "Mechanical resonant immunospecific biological detector," Appl. Phys. Lett., vol. 77, no. 3, p. 450, Jul. 2000.

110. Ilic, Y. Yang, and H. G. Craighead, "Virus detection using nanoelectromechanical devices," Appl. Phys. Lett., vol. 85, no. 13, p. 2604, 2004.

111. K. Y. Gfeller, N. Nugaeva, and M. Hegner, "Rapid biosensor for detection of antibiotic-selective growth of Escherichia coli," Appl. Environ. Microbiol., vol. 71, no. 5, pp. 2626–2631, Apr. 2005.

112. N. Nugaeva, K. Y. Gfeller, N. Backmann, H. P. Lang, M. Düggelin, and M. Hegner, "Micromechanical cantilever array sensors for selective fungal immobilization and fast growth detection," Biosensors Bioelectron., vol. 21, no. 6, pp. 849–856, Dec. 2004.

113. A. Savran, S. M. Knudsen, A. D. Ellington, and S. R. Manalis, "Micromechanical detection of proteins using aptamer-based receptor molecules," Anal. Chem., vol. 76, no. 11, pp. 3194–3198, Jun. 2004.

114. Y. Lin and S. D. Jayasena, "Inhibition of multiple thermostable DNA polymerases by a heterodimeric aptamer," J. Molecular Biol., vol. 271, no. 1, pp. 100–111, Aug. 1997.

115. Subramanian et al., "Glucose biosensing using an enzyme-coated microcantilever," Appl. Phys. Lett., vol. 81, no. 2, p. 385, Jul. 2002.

116. J. Pei, F. Tian, and T. Thundat, "Glucose biosensor based on the microcantilever," Anal. Chem., vol. 76, no. 2, pp. 292–297, Jan. 2004.

117. X. Yan, H.-F. Ji, and Y. Lvov, "Modification of microcantilevers using layer-by-layer nanoassembly film for glucose measurement," Chem. Phys. Lett., vol. 396, no. 1, pp. 34–37, Sep. 2004.

118. U. Starke et al., "The (0001)-surface of 6H-SiC: Morphology, composition and structure," App. Surf. Sci., vol. 89, no. 2, pp. 175–185, Jun. 1995.

119. S. Nishino, J. A. Powell, and H. A. Will, "Production of large-area singlecrystal wafers of cubic SiC for semiconductor devices," Appl. Phys. Lett., vol. 42, no. 5, p. 460, 1983.

120. Coletti, Silicon Carbide Biocompatibility, Surface Control, and Electronic Cellular Interaction for Biosensing Applications. Ann Arbor, MI, USA: ProQuest, 2007.

121. M. Levinshtein, S. Rumyantsev, and M. S. Shur, Special Issue: SiC Materials and Devices. Singapore: World Scientific, 2006.

122. S. E. Saddow and A. Agarwal, Advances in Silicon Carbide Processing and Applications. Norwood, MA, USA: Artech House, 2004.

123. M. E. Lin et al., "GaN grown on hydrogen plasma cleaned 6H-SiC substrates," Appl. Phys., vol. 62, no. 7, p. 702, 1993.

124. M. Mehregany, C. A. Zorman, N. Rajan, and C. H. Wu, "Silicon carbide MEMS for harsh environments," Proc. IEEE, vol. 86, no. 8, pp. 1594–1609, Aug. 1998.

125. Y. T. Yang et al., "Monocrystalline silicon carbide nanoelectromechanical systems," Appl. Phys. Lett., vol. 78, no. 2, p. 162, 2001.

126. Oliveros, A. Guiseppi-Elie, and S. E. Saddow, "Silicon carbide: A versatile material for biosensor applications," Biomed. Microdevices, vol. 15, no. 2, pp. 353–368, Apr. 2013.

127. Locke et al., "3C-SiC films on Si for MEMS applications: Mechanical properties," Mater. Sci. Forum, vols. 615–617, pp. 633–636, Mar. 2009.

128. Ellison et al., "High temperature CVD growth of SiC," Mater. Sci. Eng. B, vols. 61–62, pp. 113–120, Jul. 1998.

129. T. Seyller, "Passivation of hexagonal SiC surfaces by hydrogen termination," J. Phys., Condens. Matter, vol. 16, no. 17, pp. S1755–S1782, Apr. 2004.

130. Locke, C. Frewin, L. Abbati, and S. E. Saddow, "Demonstration of 3C-SiC MEMS structures on polysilicon-on-oxide substrates," MRS Proc., vol. 1246, p. 1246-B08-05, Feb. 2011.

131. A. Zorman et al., "Epitaxial growth of 3C–SiC films on 4 in. diam (100) silicon wafers by atmospheric pressure chemical vapor deposition," J. Appl. Phys., vol. 78, no. 8, pp. 5136–5138, Oct. 1995.

132. M. Syväjärvi et al., "Liquid phase epitaxial growth of SiC," J. Crystal Growth, vol. 197, nos. 1–2, pp. 147–154, Feb. 1998.

133. J. Chen, A. J. Steckl, and M. J. Loboda, "Molecular beam epitaxy growth of SiC on Si(111) from silacyclobutane," J. Vac. Sci. Technol. B, vol. 16, no. 3, pp. 1305–1308, May 1998.

134. M. Reyes, Y. Shishkin, S. Harvey, and S. E. Saddow, "Increased growth rates of 3C-SiC on Si (100) substrates via HCl growth additive," Mater. Sci. Forum, vols. 556–557, pp. 191–194, Sep. 2007.

135. R. Yakimova, R. M. Petoral, Jr., G. R. Yazdi, C. Vahlberg, A. L. Spetz, and K. Uvdal, "Surface functionalization and biomedical

applications based on SiC," J. Phys. D, Appl. Phys., vol. 40, no. 20, pp. 6435–6442, Oct. 2007.

136. X. Li, X. Wang, R. Bondokov, J. Morris, Y. H. An, and T. S. Sudarshan, "Micro/nanoscale mechanical and tribological characterization of SiC for orthopedic applications," J. Biomed. Mater. Res. B, Appl. Biomater., vol. 72B, no. 2, pp. 353–361, Feb. 2005.

137. L. Harris, Properties of Silicon Carbide. London: INSPEC, 1995, pp. 1–297. [137] C. H. Li, I. B. Bhat, R. Wang, and J. Seiler, "Electrochemical mechanical polishing of silicon carbide," J. Electron. Mater., vol. 33, no. 5, pp. 481–486, May 2004.

138. L. Zhou, V. Audurier, P. Pirouz, and J. A. Powell, "Chemomechanical polishing of silicon carbide," J. Electrochem. Soc., vol. 144, no. 6, pp. L161–L163, Jun. 1997.

139. C.-M. Zetterling, Process Technology for Silicon Carbide Devices. Edison, NJ, USA: IET, 2002.

140. Bernhardt, J. Schardt, U. Starke, and K. Heinz, "Epitaxially ideal oxide–semiconductor interfaces: Silicate adlayers on hexagonal (0001) and (0001)(000 ¯ 1) SiC surfaces," ¯ Appl. Phys. Lett., vol. 74, no. 8, pp. 1084–1086, 1999.

141. S. Soubatch, S. E. Saddow, S. P. Rao, W. Y. Lee, M. Konuma, and U. Starke, "Structure and morphology of 4H-SiC wafer surfaces after H2-etching," Mater. Sci. Forum, vols. 483–485, pp. 761–764, May 2005.

142. X. L. Wu, J. Y. Fan, T. Qiu, X. Yang, G. G. Siu, and P. K. Chu, "Experimental evidence for the quantum confinement effect in 3C-SiC nanocrystallites," Phys. Rev. Lett., vol. 94, no. 2, p. 026102, Jan. 2005.

143. J. Botsoa, J. M. Bluet, V. Lysenko, O. Marty, D. Barbier, and G. Guillot, "Photoluminescence of 6H-SiC nanostructures fabricated by

electrochemical etching," J. Appl. Phys., vol. 102, no. 8, p. 083526, 2007.

144. H. Lin, J. A. Gerbec, M. Sushchikh, and E. W. McFarland, "Synthesis of amorphous silicon carbide nanoparticles in a low temperature low pressure plasma reactor," Nanotechnology, vol. 19, no. 32, p. 325601, Jul. 2008.

145. Y. Leconte, M. Leparoux, X. Portier, and N. Herlin-Boime, "Controlled synthesis of β-SiC nanopowders with variable stoichiometry using inductively coupled plasma," Plasma Chem. Plasma Process., vol. 28, no. 2, pp. 233–248, Apr. 2008.

146. S. Yang, H. Zeng, H. Zhao, H. Zhang, and W. Cai, "Luminescent hollow carbon shells and fullerene-like carbon spheres produced by laser ablation with toluene," J. Mater. Chem., vol. 21, no. 12, pp. 4432–4436, 2011.

147. J. Botsoa, V. Lysenko, A. Góloën, O. Marty, J. M. Bluet, and G. Guillot, "Application of 3C-SiC quantum dots for living cell imaging," Appl. Phys. Lett., vol. 92, no. 17, p. 173902, 2008.

148. H. Williams et al., "Immobilization of streptavidin on 4H–SiC for biosensor development," Appl. Surf. Sci., vol. 258, no. 16, pp. 6056–6063, Jun. 2012.

149. S. Brahim, A. M. Wilson, D. Narinesingh, E. Iwuoha, and A. Guiseppi-Elie, "Chemical and biological sensors based on electrochemical detection using conducting electroactive polymers," Microchim. Acta, vol. 143, no. 2, pp. 123–137, Dec. 2003.

150. R. M. Petoral, Jr., G. R. Yazdi, A. L. Spetz, R. Yakimova, and K. Uvdal, "Organosilane-functionalized wide band gap semiconductor surfaces," Appl. Phys. Lett., vol. 90, no. 22, p. 223904, 2007.

151. M. Stutzmann, J. A. Garrido, M. Eickhoff, and M. S. Brandt, "Direct biofunctionalization of semiconductors: A survey," Phys. Status Solidi A, vol. 203, no. 14, pp. 3424–3437, Nov. 2006.

152. T. Matsuo, M. Esashi, and H. Abe, "ISFET's using inorganic gate thin films," IEEE Trans. Electron Devices, vol. 26, no. 12, pp. 1939–1944, Dec. 1979.

153. Jakobson, I. Bloom, and Y. Nemirovsky, "1/f noise in CMOS transistors for analog applications from subthreshold to saturation," Solid-State Electron., vol. 42, no. 10, pp. 1807–1817, Oct. 1998.

154. [154] A. L. Spetz et al., "New transducer material concepts for biosensors and surface functionalization," Proc. SPIE, vol. 7362, p. 736206, May 2009.

155. S. J. Schoell et al., "Functionalization of 6H-SiC surfaces with organosilanes," Appl. Phys. Lett., vol. 92, no. 15, p. 153301, 2008.

156. S. Dhar, O. Seitz, M. D. Halls, S. Choi, Y. J. Chabal, and L. C. Feldman, "Chemical properties of oxidized silicon carbide surfaces upon etching in hydrofluoric acid," J. Amer. Chem. Soc., vol. 131, no. 46, pp. 16808–16813, Nov. 2009.

157. M. Preuss, F. Bechstedt, W. G. Schmidt, Jr., J. Sochos, B. Schröter, and W. Richter, "Clean and pyrrole-functionalized Si- and C-terminated SiC surfaces: First-principles calculations of geometry and energetics compared with LEED and XPS," Phys. Rev. B, vol. 74, no. 2, p. 235406, Dec. 2006.

158. Catellani and G. Cicero, "Modifications of cubic SiC surfaces studied by ab initio simulations: From gas adsorption to organic functionalization," J. Phys. D, Appl. Phys., vol. 40, no. 20, pp. 6215–6224, Oct. 2007.

159. G. Cicero and A. Catellani, "Towards SiC surface functionalization: An ab initio study," J. Chem. Phys., vol. 122, no. 21, p. 214716, 2005.

160. P. Godignon, "SiC materials and technologies for sensors development," Mater. Sci. Forum, vols. 483–485, pp. 1009–1014, May 2005.

161. P. Godignon, I. Martin, G. Gabriel, R. Gömez, M. Placidi, and R. Villa, "New generation of SiC based biodevices implemented on 4' wafers," Mater. Sci. Forum, vols. 645–648, pp. 1097–1100, Apr. 2010.

162. H. O. Pierson, Handbook of Chemical Vapor Deposition, 2nd ed. Norwich, NY, USA: William Andrew, 1999.

163. G. Gabriel et al., "Manufacturing and full characterization of silicon carbide-based multi-sensor micro-probes for biomedical applications," Microelectron. J., vol. 38, no. 3, pp. 406–415, Mar. 2007.

164. H. Tsuchida, I. Kamata, and K. Izumi, "Infrared attenuated total reflection spectroscopy of 6H-SiC(0001) and (000(1)over-bar) surfaces," J. Appl. Phys., vol. 85, no. 7, pp. 3569–3575, 1999.

165. N. G. Wright and A. B. Horsfall, "SiC sensors: A review," J. Phys. D, Appl. Phys., vol. 40, no. 20, pp. 6345–6354, 2007.

166. N. Marsi, B. Y. Majlis, A. A. Hamzah, and F. Mohd-Yasin, "The mechanical and electrical effects of MEMS capacitive pressure sensor based 3C-SiC for extreme temperature," J. Eng., vol. 2014, May 2014, Art. ID 715167.

167. M. J. Madou, Fundamentals of Microfabrication. Boca Raton, FL, USA: CRC Press, 2002.

168. J. Mertens, M. Álvarez, and J. Tamayo, "Real-time profile of microcantilevers for sensing applications," Appl. Phys. Lett., vol. 87, no. 23, p. 234102, 2005.

169. S. Chatzandroulis, A. Tserepi, D. Goustouridis, P. Normand, and D. Tsoukalas, "Fabrication of single crystal Si cantilevers using a dry release process and application in a capacitive-type humidity sensor," Microelectron. Eng., vols. 61–62, pp. 955–961, Jul. 2002.

170. R. Kermany, G. Brawley, N. Mishra, E. Sheridan, W. P. Bowen, and F. Iacopi, "Microresonators with Q-factors over a million from

highly stressed epitaxial silicon carbide on silicon," Appl. Phys. Lett., vol. 104, no. 8, p. 081901, Feb. 2014.

171. Iacopi, R. E. Brock, A. Iacopi, L. Hold, and R. H. Dauskardt, "Evidence of a highly compressed nanolayer at the epitaxial silicon carbide interface with silicon," Acta Mater., vol. 61, no. 17, pp. 6533–6540, Oct. 2013.

172. L. Sekaric, J. M. Parpia, H. G. Craighead, T. Feygelson, B. H. Houston, and J. E. Butler, "Nanomechanical resonant structures in nanocrystalline diamond," Appl. Phys. Lett., vol. 81, no. 23, p. 4455, 2002.

173. M. Placidi et al., "Fabrication of monocrystalline 3C–SiC resonators for MHz frequency sensors applications," Sens. Actuators B, Chem., vol. 133, no. 1, pp. 276–280, Jul. 2008.

174. S. Schmid, Jr., K. D. Jensen, K. H. Nielsen, and A. Boisen, "Damping mechanisms in high-Q micro and nanomechanical string resonators," Phys. Rev. B, vol. 84, no. 1, p. 165307, Oct. 2011.

175. M. Houmad, A. Abbassi, A. Benyoussef, H. Ez-Zahraouy, and A. El Kenz, "Optical properties of Ni doped 3C-SiC with ab initio calculations," presented at the Int. Renew. Sustain. Energy Conf. (IRSEC), Oct. 2014, pp. 596–601.

176. X. Wang and K. M. Liew, "Silicon carbide nanotubes serving as a highly sensitive gas chemical sensor for formaldehyde," J. Phys. Chem. C, vol. 115, no. 21, pp. 10388–10393, Jun. 2011.

177. M. Wijesundara and R. Azevedo, Silicon Carbide Microsystems for Harsh Environments. Springer-Verlag New York, 2011.

178. V. Cunning, M. Ahmed, N. Mishra, A. R. Kermany, B. Wood, and F. Iacopi, "Graphitized silicon carbide microbeams: Wafer-level, selfaligned graphene on silicon wafers," Nanotechnology, vol. 25, no. 32, p. 325301, Jul. 2014.

179. Gupta, M. Notarianni, N. Mishra, M. Shafiei, F. Iacopi, and N. Motta, "Evolution of epitaxial graphene layers on 3C SiC/Si (111) as

a function of annealing temperature in UHV," Carbon, vol. 68, pp. 563–572, Mar. 2014.

180. Gupta, E. Placidi, C. Hogan, N. Mishra, F. Iacopi, and N. Motta, "The transition from 3C SiC (111) to graphene captured by ultra high vacuum scanning tunneling microscopy," Carbon, vol. 91, pp. 378–385, Sep. 2015.

181. V. Singh, D. Joung, L. Zhai, S. Das, S. I. Khondaker, and S. Seal, "Graphene based materials: Past, present and future," Prog. Mater. Sci., vol. 56, no. 8, pp. 1178–1271, Oct. 2011.

182. K. S. Novoselov et al., "Electric field effect in atomically thin carbon films," Science, vol. 306, no. 5696, pp. 666–669, Oct. 2004.

183. K. S. Novoselov et al., "Room-temperature quantum Hall effect in graphene," Science, vol. 315, no. 5817, p. 1379, Mar. 2007.

184. Y. Zhang, Y.-W. Tan, H. L. Stormer, and P. Kim, "Experimental observation of the quantum Hall effect and Berry's phase in graphene," Nature, vol. 438, no. 7, pp. 201–204, Nov. 2005.

185. S. V. Morozov et al., "Strong suppression of weak localization in graphene," Phys. Rev. Lett., vol. 97, no. 1, p. 016801, Jul. 2006.

186. W. Choi, I. Lahiri, R. Seelaboyina, and Y. S. Kang, "Synthesis of graphene and its applications: A review," Critical Rev. Solid State Mater. Sci., vol. 35, no. 1, pp. 52–71, Feb. 2010.

187. A. Balandin et al., "Superior thermal conductivity of single-layer graphene," Nano Lett., vol. 8, no. 3, pp. 902–907, 2008.

188. J.-H. Chen, C. Jang, S. Xiao, M. Ishigami, and M. S. Fuhrer, "Intrinsic and extrinsic performance limits of graphene devices on SiO2," Nature Nanotechnol., vol. 3, no. 4, pp. 206–209, Mar. 2008.

189. N. O. Weiss et al., "Graphene: An emerging electronic material," Adv. Mater., vol. 24, no. 43, pp. 5782–5825, Nov. 2012.

190. K. S. Novoselov et al., "Two-dimensional gas of massless Dirac fermions in graphene," Nature, vol. 438, no. 7065, pp. 197–200, Nov. 2005.

191. Q. Bao and K. P. Loh, "Graphene photonics, plasmonics, and broadband optoelectronic devices," ACS Nano, vol. 6, no. 5, pp. 3677–3694, May 2012.

192. W. C. Lee et al., "Origin of enhanced stem cell growth and differentiation on graphene and graphene oxide," ACS Nano, vol. 5, no. 9, pp. 7334–7341, Sep. 2011.

193. P. Orlita et al., "Approaching the Dirac point in high-mobility multilayer epitaxial graphene," Phys. Rev. Lett., vol. 101, p. 267601, Dec. 2008.

194. Lee, X. Wei, J. W. Kysar, and J. Hone, "Measurement of the elastic properties and intrinsic strength of monolayer graphene," Science, vol. 321, no. 5887, pp. 385–388, Jul. 2008.

195. X. Li, X. Wang, L. Zhang, S. Lee, and H. Dai, "Chemically derived, ultrasmooth graphene nanoribbon semiconductors," Science, vol. 319, no. 5867, pp. 1229–1232, Feb. 2008.

196. K. A. Ritter and J. W. Lyding, "The influence of edge structure on the electronic properties of graphene quantum dots and nanoribbons," Nature Mater., vol. 8, no. 3, pp. 235–242, Feb. 2009.

197. [197] A. M. Pinto, I. C. Gonçalves, and F. D. Magalhães, "Graphene-based materials biocompatibility: A review," Colloids Surf. B, Biointerfaces, vol. 111, pp. 188–202, Nov. 2013.

198. V. Dhand, K. Y. Rhee, H. J. Kim, and D. H. Jung, "A comprehensive review of graphene nanocomposites: Research status and trends," J. Nanomater., vol. 2013, Jan. 2013, Art. ID 158.

199. K.-H. Liao, Y.-S. Lin, C. W. Macosko, and C. L. Haynes, "Cytotoxicity of graphene oxide and graphene in human erythrocytes and skin fibroblasts," ACS Appl. Mater. Interfaces, vol. 3, no. 7, pp. 2607–2615, Jun. 2011.

200. [200] S. Liu et al., "Antibacterial activity of graphite, graphite oxide, graphene oxide, and reduced graphene oxide: Membrane and oxidative stress," ACS Nano, vol. 5, no. 9, pp. 6971–6980, Sep. 2011.

201. R. M. Donlan, "Biofilms and device-associated infections," Emerg. Infectious Diseases, vol. 7, no. 2, pp. 277–281, Mar. 2001.

202. H. Anwar, J. L. Strap, K. Chen, and J. W. Costerton, "Dynamic interactions of biofilms of mucoid Pseudomonas aeruginosa with tobramycin and piperacillin," Antimicrob. Agents. Chemother., vol. 36, no. 6, pp. 1208–1214, May 1992.

203. M. Gabriel, K. Nazmi, E. C. Veerman, A. V. N. Amerongen, and A. Zentner, "Preparation of LL-37-grafted titanium surfaces with bactericidal activity," Bioconjugate Chem., vol. 17, no. 2, pp. 548–550, Jan. 2006.

204. M. Santos, M. C. R. Tria, R. A. M. V. Vergara, F. Ahmed, R. C. Advincula, and D. F. Rodrigues, "Antimicrobial graphene polymer (PVK-GO) nanocomposite films," Chem. Commun., vol. 47, no. 31, pp. 8892–8894, Aug. 2011.

205. E. M. Carpio, C. M. Santos, X. Wei, and D. F. Rodrigues, "Toxicity of a polymer–graphene oxide composite against bacterial planktonic cells, biofilms, and mammalian cells," Nanoscale, vol. 4, no. 5, pp. 4746–4756, Aug. 2012.

206. Peng et al., "Synthesis of manganese ferrite/graphene oxide nanocomposites for biomedical applications," Small, vol. 8, no. 23, pp. 3620–3630, Dec. 2012.

207. W. C. Lee, C. H. Lim, C. S. Kenry, K. P. Loh, and C. T. Lim, "Cellassembled graphene biocomposite for enhanced chondrogenic differentiation," Small, vol. 11, no. 8, pp. 963–969, Jan. 2015.

208. Reina et al., "Large area, few-layer graphene films on arbitrary substrates by chemical vapor deposition," Nano Lett., vol. 9, no. 1, pp. 30–35, Dec. 2008.

209. K. Geim and K. S. Novoselov, "The rise of graphene," Nature Mater., vol. 6, no. 3, pp. 183–191, Mar. 2007.

210. P. R. Somani, S. P. Somani, and M. Umeno, "Planer nano-graphenes from camphor by CVD," Chem. Phys. Lett., vol. 430, nos. 1–3, pp. 56–59, 2006.

211. J. Wang, M. Zhu, R. A. Outlaw, X. Zhao, D. M. Manos, and B. C. Holloway, "Synthesis of carbon nanosheets by inductively coupled radio-frequency plasma enhanced chemical vapor deposition," Carbon, vol. 42, no. 14, pp. 2867–2872, 2004.

212. N. Obraztsov, A. A. Zolotukhin, A. O. Ustinov, A. P. Volkov, Y. Svirko, and K. Jefimovs, "DC discharge plasma studies for nanostructured carbon CVD," Diamond Rel. Mater., vol. 12, nos. 3–7, pp. 917–920, 2003.

213. S. Stankovich et al., "Synthesis of graphene-based nanosheets via chemical reduction of exfoliated graphite oxide," Carbon, vol. 45, no. 7, pp. 1558–1565, 2007.

214. S. Gilje, S. Han, M. Wang, K. L. Wang, and R. B. Kaner, "A chemical route to graphene for device applications," Nano Lett., vol. 7, no. 11, pp. 3394–3398, Nov. 2007.

215. V. Kosynkin et al., "Longitudinal unzipping of carbon nanotubes to form graphene nanoribbons," Nature, vol. 458, no. 7240, pp. 872–876, Apr. 2009.

216. Santos, L. Chico, and L. Brey, "Carbon nanoelectronics: Unzipping tubes into graphene ribbons," Phys. Rev. Lett., vol. 103, p. 086801, Aug. 2009.

217. L. Jiao, L. Zhang, X. Wang, G. Diankov, and H. Dai, "Narrow graphene nanoribbons from carbon nanotubes," Nature, vol. 458, no. 7240, pp. 877–880, Apr. 2009.

218. Sinitskii, A. Dimiev, D. V. Kosynkin, and J. M. Tour, "Graphene nanoribbon devices produced by oxidative unzipping of carbon nanotubes," ACS Nano, vol. 4, no. 9, pp. 5405–5413, Sep. 2010.

219. Kageshima, H. Hibino, and S. Tanabe, "The physics of epitaxial graphene on SiC(0001)," J. Phys., Condens. Matter, vol. 24, no. 31, p. 314215, 2012.

220. Ouerghi et al., "Large-area and high-quality epitaxial graphene on off-axis SiC wafers," ACS Nano, vol. 6, no. 7, pp. 6075–6082, Jun. 2012.

221. Eda, G. Fanchini, and M. Chhowalla, "Large-area ultrathin films of reduced graphene oxide as a transparent and flexible electronic material," Nature Nanotechnol., vol. 3, no. 5, pp. 270–274, Apr. 2008

222. A. Becerril, J. Mao, Z. Liu, R. M. Stoltenberg, Z. Bao, and Y. Chen, "Evaluation of solution-processed reduced graphene oxide films as transparent conductors," ACS Nano, vol. 2, no. 3, pp. 463–470, Mar. 2008.

223. Y.-W. Son, M. L. Cohen, and S. G. Louie, "Energy gaps in graphene nanoribbons," Phys. Rev. Lett., vol. 97, no. 21, p. 216803, Nov. 2006.

224. V. Barone, O. Hod, and G. E. Scuseria, "Electronic structure and stability of semiconducting graphene nanoribbons," Nano Lett., vol. 6, no. 12, pp. 2748–2754, Dec. 2006.

225. T. Shimizu et al., "Large intrinsic energy bandgaps in annealed nanotubederived graphene nanoribbons," Nature Nanotechnol., vol. 6, no. 1, pp. 45–50, Dec. 2010.

226. M. Y. Han, B. Özyilmaz, Y. Zhang, and P. Kim, "Energy band-gap engineering of graphene nanoribbons," Phys. Rev. Lett., vol. 98, no. 20, p. 206805, May 2007.

227. L. Elías et al., "Longitudinal cutting of pure and doped carbon nanotubes to form graphitic nanoribbons using metal clusters as nanoscalpels," Nano Lett., vol. 10, no. 2, pp. 366–372, Feb. 2010.

228. L. Tedesco et al., "Hall effect mobility of epitaxial graphene grown on silicon carbide," Appl. Phys. Lett., vol. 95, no. 12, p. 122102, 2009.

229. P. Xu, H. Yu, X. Xia, F. Yu, M. Liu, and X. Li, "Resonant cantilevers with nanoparticles-spaced functional graphene-oxide sheets for highperformance sensing to ppt-level explosive vapor," in Proc. IEEE 26th Int. Conf. Micro Electro Mech. Syst., Jan. 2013, pp. 989–992.

230. Kang et al., "Efficient transfer of large-area graphene films onto rigid substrates by hot pressing," ACS Nano, vol. 6, no. 6, pp. 5360–5365, Jun. 2012.

231. J. D. Caldwell et al., "Technique for the dry transfer of epitaxial graphene onto arbitrary substrates," ACS Nano, vol. 4, no. 2, pp. 1108–1114, Oct. 2009. [Online]. Available: http://arxiv.org/abs/0910.2624

232. Iacopi et al., "A catalytic alloy approach for graphene on epitaxial SiC on silicon wafers," J. Mater. Res., vol. 30, no. 5, pp. 609–616, Feb. 2015.

233. [233] P. Macháč, T. Fidler, S. Cichoň, and V. Jurka, "Synthesis of graphene on Co/SiC structure," J. Mater. Sci., Mater. Electron., vol. 24, no. 10, pp. 3793–3799, 2013.

234. 4] T. Yoneda, M. Shibuya, K. Mitsuhara, A. Visikovskiy, Y. Hoshino, and Y. Kido, "Graphene on SiC(0001) and SiC(0001) surfaces grown via ¯ Ni-silicidation reactions," Surf. Sci., vol. 604, nos. 17–18, pp. 1509–1515, Aug. 2010.

235. J. Van Bommel, J. E. Crombeen, and A. Van Tooren, "LEED and Auger electron observations of the SiC(0001) surface," Surf. Sci., vol. 48, no. 2, pp. 463–472, Mar. 1975.

236. Rollings et al., "Synthesis and characterization of atomically thin graphite films on a silicon carbide substrate," J. Phys. Chem. Solids, vol. 67, nos. 9–10, pp. 2172–2177, Sep./Oct. 2006.

237. Z.-Y. Juang et al., "Synthesis of graphene on silicon carbide substrates at low temperature," Carbon, vol. 47, no. 8, pp. 2026–2031, 2009.

238. Escobedo-Cousin et al., "Local solid phase growth of few-layer graphene on silicon carbide from nickel silicide supersaturated with carbon," J. Appl. Phys., vol. 113, no. 11, p. 114309, Mar. 2013.

239. Berger et al., "Ultrathin epitaxial graphite: 2D electron gas properties and a route toward graphene-based nanoelectronics," J. Phys. Chem. B, vol. 108, no. 52, pp. 19912–19916, Dec. 2004.

240. Berger et al., "Electronic confinement and coherence in patterned epitaxial graphene," Science, vol. 312, no. 5777, pp. 1191–1196, May 2006.

241. J. Penuelas et al., "Surface morphology and characterization of thin graphene films on SiC vicinal substrate," Phys. Rev. B, vol. 79, no. 3, p. 033408, Jan. 2009.

242. V. Emtsev et al., "Towards wafer-size graphene layers by atmospheric pressure graphitization of silicon carbide," Nature Mater., vol. 8, no. 3, pp. 203–207, Feb. 2009.

243. T. Ohta, N. C. Bartelt, S. Nie, K. Thürmer, and G. L. Kellogg, "Role of carbon surface diffusion on the growth of epitaxial graphene on SiC," Phys. Rev. B, vol. 81, no. 12, p. 121411(R), Mar. 2010.

244. Lauwers et al., "Materials aspects, electrical performance, and scalability of Ni silicide towards sub-0.13 μm technologies," J. Vac. Sci. Technol. B, vol. 19, no. 6, p. 2026, 2001.

245. P. Macháč, T. Fidler, S. Cichoň, and L. Mišková, "Synthesis of graphene on SiC substrate via Ni-silicidation reactions," Thin Solid Films, vol. 520, no. 16, pp. 5215–5218, 2012.

246. Iacopi et al., "Orientation-dependent stress relaxation in heteroepitaxial 3C-SiC films," Appl. Phys. Lett., vol. 102, no. 1, p. 011908, 2013.

247. V. Y. Aristov et al., "Graphene synthesis on cubic SiC/Si wafers. perspectives for mass production of graphene-based electronic devices," Nano Lett., vol. 10, no. 3, pp. 992–995, Mar. 2010.

248. U. Starke, C. Coletti, K. Emtsev, A. A. Zakharov, T. Ouisse, and D. Chaussende, "Large area quasi-free standing monolayer graphene on 3C-SiC(111)," Mater. Sci., vols. 717–720, pp. 617–620, May 2012.

249. P. Sutter, "Epitaxial graphene: How silicon leaves the scene," Nature Mater., vol. 8, no. 3, pp. 171–172, Feb. 2009.

250. R. Takahashi et al., "Low-energy-electron-diffraction and X-rayphototelectron-spectroscopy studies of graphitization of 3C-SiC(111) thin film on Si(111) substrate," Jpn. J. Appl. Phys., vol. 50, no. 7R, p. 070103, 2011.

251. Bostwick et al., "Symmetry breaking in few layer graphene films," New J. Phys., vol. 9, no. 1, p. 385, Oct. 2007.

252. Ouerghi et al., "Structural coherency of epitaxial graphene on 3C-SiC(111) epilayers on Si(111)," Appl. Phys. Lett., vol. 97, no. 16, p. 161905, Oct. 2010.

253. Ouerghi et al., "Sharp interface in epitaxial graphene layers on 3C-SiC(100)/Si(100) wafers," Phys. Rev. B, vol. 83, no. 20, p. 205429, 2011.

254. K. V. Emtsev, F. Speck, T. Seyller, L. Ley, and J. D. Riley, "Interaction, growth, and ordering of epitaxial graphene on SiC{0001} surfaces: A comparative photoelectron spectroscopy study," Phys. Rev. B, vol. 77, no. 15, p. 155303, Apr. 2008.

255. Z. H. Ni et al., "Raman spectroscopy of epitaxial graphene on a SiC substrate," Phys. Rev. B, vol. 77, p. 115416, Mar. 2008.

256. Kästle, H. G. Boyen, A. Schröder, A. Plettl, and P. Ziemann, "Size effect of the resistivity of thin epitaxial gold films," Phys. Rev. B, vol. 70, no. 1, p. 165414, Oct. 2004.

257. T. Kuila, S. Bose, A. K. Mishra, P. Khanra, N. H. Kim, and J. H. Lee, "Chemical functionalization of graphene and its applications," Prog. Mater. Sci., vol. 57, no. 7, pp. 1061–1105, Sep. 2012.

258. Fang, K. Wang, H. Lu, Y. Yang, and S. Nutt, "Single-layer graphene nanosheets with controlled grafting of polymer chains," J. Mater. Chem., vol. 20, no. 10, pp. 1982–1992, 2010.

259. [259] M. J. Park, J. K. Lee, B. S. Lee, Y.-W. Lee, I. S. Choi, and S.-G. Lee, "Covalent modification of multiwalled carbon nanotubes with imidazolium-based ionic liquids: Effect of anions on solubility," Chem. Mater., vol. 18, no. 6, pp. 1546–1551, 2006.

260. [260] H. Yang et al., "Covalent functionalization of chemically converted graphene sheets via silane and its reinforcement," J. Mater. Chem., vol. 19, no. 26, pp. 4632–4638, 2009.

261. B. Bourlinos, D. Gournis, D. Petridis, T. Szabó, A. Szeri, and I. Dékány, "Graphite oxide: Chemical reduction to graphite and surface modification with primary aliphatic amines and amino acids," Langmuir, vol. 19, no. 15, pp. 6050–6055, 2003.

262. Bekyarova et al., "Chemical modification of epitaxial graphene: Spontaneous grafting of aryl groups," J. Amer. Chem. Soc., vol. 131, no. 4, pp. 1336–1337, Feb. 2009.

263. Y. Zhu, A. L. Higginbotham, and J. M. Tour, "Covalent functionalization of surfactant-wrapped graphene nanoribbons," Chem. Mater., vol. 21, no. 21, pp. 5284–5291, 2009.

264. S. Stankovich, R. D. Piner, S. T. Nguyen, and R. S. Ruoff, "Synthesis and exfoliation of isocyanate-treated graphene oxide nanoplatelets," Carbon, vol. 44, no. 15, pp. 3342–3347, Dec. 2006.

265. Z. Liu, J. T. Robinson, X. Sun, and H. Dai, "PEGylated nanographene oxide for delivery of water-insoluble cancer drugs," J. Amer. Chem. Soc., vol. 130, no. 33, pp. 10876–10877, Aug. 2008.

266. S. Vadukumpully, J. Gupta, Y. Zhang, G. Q. Xu, and S. Valiyaveettil, "Functionalization of surfactant wrapped graphene nanosheets with alkylazides for enhanced dispersibility," Nanoscale, vol. 3, no. 1, pp. 303–308, Jan. 2011. [

267. X. Xu et al., "Functionalization of graphene sheets by polyacetylene: Convenient synthesis and enhanced emission," Macromolecular Chem. Phys., vol. 212, no. 8, pp. 768–773, 2011.

268. T. Kuila et al., "Preparation of functionalized graphene/linear low density polyethylene composites by a solution mixing method," Carbon, vol. 49, no. 3, pp. 1033–1037, Mar. 2011.

269. Shan, H. Yang, D. Han, Q. Zhang, A. Ivaska, and L. Niu, "Water-soluble graphene covalently functionalized by biocompatible poly-L-lysine," Langmuir, vol. 25, no. 20, pp. 12030–12033, Oct. 2009.

270. D.-D. Zhang, S.-Z. Zu, and B.-H. Han, "Inorganic–organic hybrid porous materials based on graphite oxide sheets," Carbon, vol. 47, no. 13, pp. 2993–3000, Nov. 2009.

271. K. Yang, L. Feng, H. Hong, W. Cai, and Z. Liu, "Preparation and functionalization of graphene nanocomposites for biomedical applications," Nature Protocols, vol. 8, no. 12, pp. 2392–2403, Nov. 2013.

3

ETCHING OF SILICON CARBIDE USING CHLORINE TRIFLUORIDE GAS

Hitoshi Habuka[1]

[1] Department of Chemical and Energy Engineering, Yokohama National University, Japan

1. INTRODUCTION

Silicon carbide (SiC) is known as an excellent material. Single-crystalline 4H-silicon carbide is a fascinating wide band-gap semiconductor material [1-3], suitable for high power and high temperature electronic devices [4] because of its suitable properties, such as high electron mobility, high thermal conductivity, high chemical stability, high mechanical hardness, high break down electric field and small dielectric constant [4, 5]. Additionally, many researchers have reported the stability of silicon carbide micro-electromechanical systems (MEMS) under corrosive conditions using acid and alkaline chemical reagents [6-9]. Polycrystalline 3C-silicon carbide is widely used for various purposes, such as dummy wafers and reactor parts, in silicon semiconductor device production processes.

In the semiconductor materials production technology [10], the electronics devices manufacturing process needs an easy and cost effective technique, such as wet and/or dry cleaning, for preparing the clean surface of the substrate materials. However, the suitable properties of silicon carbide often provided difficult Problems. The chemically and mechanically stable nature often makes it very difficult to prepare the entire surface in the wafer production process, such as surface polishing and removal of any damaged layer. Useful chemical reagents and processes should be developed for silicon carbide material production.

Wet and dry etching methods of silicon carbide have been studied by many researchers [5, 11-21], using various gases and various wet etchants. However, the largest etching rate reported was nearly 1 μm/min. Here, chlorine trifluoride (ClF_3) gas is very reactive even at low temperatures and has a very strong capability to etch various materials, such as silicon [22] without plasma assistance.

In Section 2, details of polycrystalline 3C-silicon carbide etching using chlorine trifluoride gas [23, 24] are reviewed, particularly focusing on the etching rate, gaseous products, surface chemical bonds and the surface morphology of the silicon carbide. In Section 3, the dry etching of single-crystalline 4H-silicon carbide using chlorine trifluoride gas [25-29] over the wide temperature range of 570-1570 K is reviewed, particularly about the etching rate, surface chemical reaction rate constant, surface morphology and etch pits.

2. POLYCRYSTALLINE 3C-SILICON CARBIDE ETCHING USING CHLORINE TRIFLUORIDE GAS

2.1. Reactor and Processes Using Chlorine Trifluoride Gas

In order to etch silicon carbide by chlorine trifluoride gas, the horizontal cold-wall reactor shown inFigure 1. is used. This reactor consists of a gas supply system, a quartz chamber and infrared lamps. A 30 mm wide x 40 mm long x 0.2-1 mm thick 3C-silicon carbide substrate manufactured using chemical vapor deposition (CVD) (Admap Inc., Tokyo) is held horizontally on the bottom wall of the quartz chamber.

The gas supply system introduces chlorine trifluoride gas, nitrogen gas and hydrogen gas. Hydrogen gas is used to remove the silicon oxide film on the silicon carbide substrate surface, the same as those on the silicon surface [22]. The height and width of the quartz chamber are compactly designed to be 10 mm and 40 mm, respectively, similar to the chamber in our various studies [22, 30].

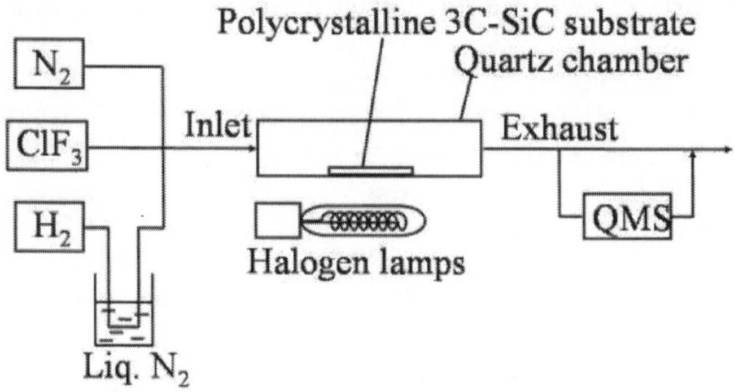

Figure 1. Horizontal cold-wall reactor used for etching polycrystalline 3C-silicon carbide substrate.

The etching using chlorine trifluoride gas is carried out following the process shown in Figure 2. There are mainly three steps.

a) cleaning the silicon carbide substrate surface by baking in ambient hydrogen at 1370 K for 10 min,

b) changing the gas from hydrogen to nitrogen, and

c) etching the silicon carbide substrate surface using chlorine trifluoride gas.

During step (a), hydrogen gas is introduced at atmospheric pressure into the reactor at a flow rate of 2 slm. Next, in step (b), the quartz chamber and the silicon carbide substrate are cooled to room temperature. The hydrogen gas present in the quartz chamber must be sufficiently purged with nitrogen gas to avoid an explosive reaction between hydrogen and chlorine trifluoride. During step (c), the silicon carbide substrate is heated and adjusted to temperatures between 670 K and 970 K. The silicon carbide substrate is etched by chlorine trifluoride (>99.9 %, Kanto Denka Kogyo Co., Ltd., Tokyo) at a flow rate of 0.1-0.25 slm without further purification and without dilution. In order to evaluate the gaseous products, part of the exhaust gas from the reactor is fed into a quadrupole mass spectra (QMS) analyzer, as shown in Figure 1. The silicon carbide etching rate using chlorine trifluoride gas is evaluated from the decrease in the weight of the silicon carbide substrate. The surface morphology of the polycrystalline 3C-silicon carbide substrate before and after the etching is observed using an optical microscope. The root-mean-square (RMS) surface roughness and the average roughness, R_a, are measured. In order to evaluate the condition of the chemical bonds of the silicon carbide surface before and after the etching, X-ray photoelectron spectra (XPS) are obtained.

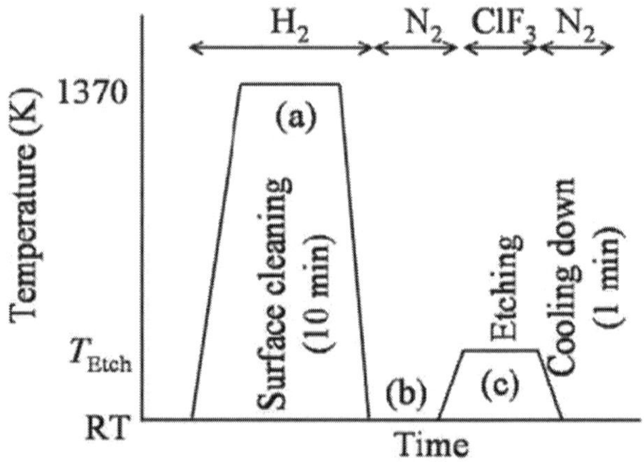

Figure 2. Process for cleaning and etching polycrystalline 3C-silicon carbide surface.

2.2. Etching Rate

The etching rate of the polycrystalline 3C-silicon carbide substrate surface is shown in Figure 3, which was obtained using 100% chlorine trifluoride gas at various gas flow rates in the temperature range of 670 to 970 K at atmospheric pressure. The squares, circles and triangles show the etching rate at the chlorine trifluoride gas flow rate of 0.2, 0.1 and 0.05 slm, respectively. As shown in Figure 3, the etching rate at the substrate temperature less than 670K is quite low; its value is less than 1 μm/min. However, with the increasing substrate temperature, the etching rate significantly increases particularly near 720 K. At the substrate temperature of 770 K, the etching rate at the flow rate of 0.2 slm becomes 20 μm/min; it remains constant at substrate temperatures greater than 770 K.

As shown in Figure 3, the etching rate changes with the flow rates. The etching rates are 25, 10 and 5 μm/min at the flow rates of 0.2, 0.1 and

0.05 slm, respectively. For each chlorine trifluoride gas flow rate, the trend in the flat etching rate at temperature greater than 770K is maintained.

In order to evaluate the influence of chlorine trifluoride gas concentration, the etching rate of the polycrystalline 3C-silicon carbide substrate surface using 10-100% chlorine trifluoride gas in ambient nitrogen was measured at the flow rate of 0.2 slm, atmospheric pressure and 670-970 K, as shown inFigure 4. In this figure, the square, reverse triangle, circle, diamond, and triangle show the substrate temperatures of 670, 730, 770, 870 and 970 K, respectively. Being consistent with Figure 3, the etching rate at the substrate temperature of 670K is very low. The etching rates at 730 K are significantly higher that those at 670 K.

At the substrate temperature of 770 K, the etching rate is proportional to the chlorine trifluoride gas concentration. At 870 and 970 K, the etching rate at each chlorine trifluoride gas concentration is the same as that at 770 K. Therefore, when the substrate temperature is higher than 770 K, the etching rate over a very wide chlorine trifluoride gas concentration range is not affected by the substrate temperature.

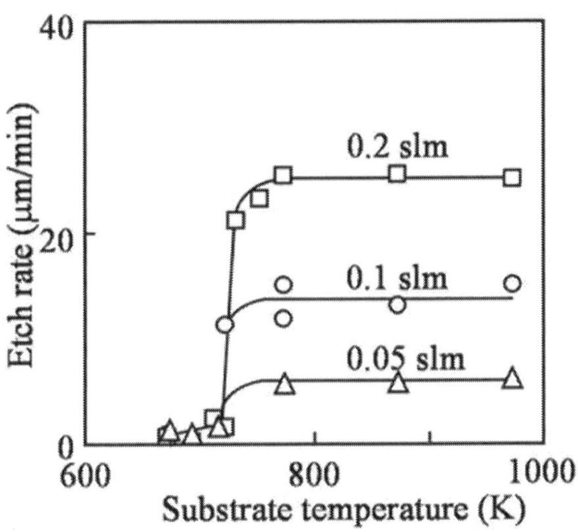

Figure 3. Etching rate of the polycrystalline 3C-silicon carbide substrate surface by chlorine trifluoride gas (100%) at atmospheric pressure in the temperature range between 670 and 970 K. Square: 0.2 slm, circle: 0.1 slm, and triangle: 0.05 slm.

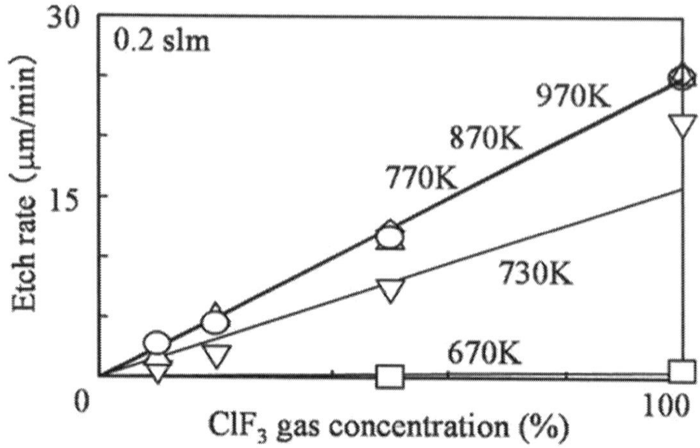

Figure 4. Etching rate of the polycrystalline 3C-silicon carbide substrate surface using chlorine trifluoride gas at 10-100%, 0.2 slm, atmospheric pressure, and 670-970K. Square: 670 K, reverse triangle: 730K, circle: 770K, diamond: 870 K, and triangle:970K.

2.3. Surface Morphology and Roughness

The change in the surface morphology of the 3C-silicon carbide is explained. Figure 5 shows photographs of the silicon carbide surface etched using chlorine trifluoride gas at the flow rate of 0.1 slm and atmospheric pressure at 670, 720 and 770 K for 15 min. The values in parentheses are the etch depth. Figure 5 (a) shows the initial surface which has the periodically line-shaped hills and valleys. As shown in Figure 5 (b), although the surface still has the line-shaped morphology at 670 K after 15 minutes, its pattern remains but is unclear. At the higher temperatures of 720 and 770 K, the line-shaped appearance does not remain, as shown in Figures 5 (c) and (d). In these figures, there are very small and very shallow pits having a round edge. This shows that the etching using chlorine trifluoride can smooth the large hills and valleys which existed on the silicon carbide surface.

In order to show the detail of surface smoothing effect, the change in the surface appearance is shown, in Figure 6, along etch period at a substrate temperature of 770 K and a flow rate of 0.1 slm of chlorine trifluoride. This figure shows photographs of the etched silicon carbide surface at (a) 0 min, (b) 5 min, (c) 10 min, (d) 15 min, and (e) 30 min. The values in parenthes are the etch depth. Figures 6 (a) and (d)are the same as Figures 5 (a) and (d), respectively. The relationship between the etch period and the etch depth is shown in Figure 7 (a). The line-shaped pattern in Figure 6 (a) is slightly rounded after 5 minutes. At 10 minutes, there is only a trace of the line-shaped appearance, as shown in Figure 6 (c). The 3C-silicon carbide surface has a round-shaped morphology after 15 minutes as shown in Figure 6 (d), since the line-shaped pattern is removed during the etch period between 10 and 15 minutes. The surface morphology in Figure 6 (d) is maintained at 30 minutes in Figure 6 (e), and the rounded edges of the very shallow pits do not become sharp during the last 15 minutes. Thus, the round-shaped morphology of silicon carbide surface will be maintained during etching for longer than 30 minutes.

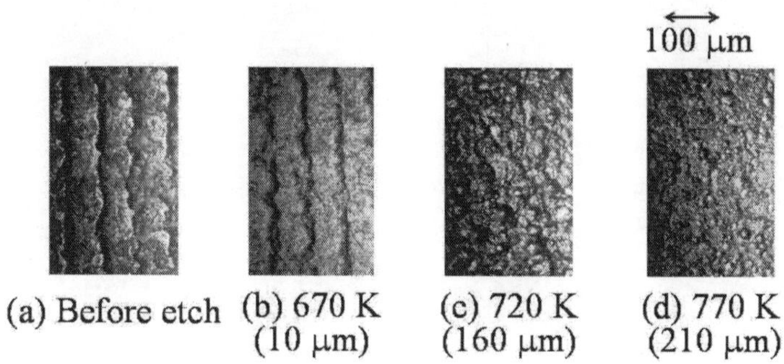

Figure 5. Photograph of the silicon carbide substrate surface etched using chlorine trifluoride at atmospheric pressure after 15 min at (b) 670 K, (c) 720 K, and (d) 770 K. (a) is the silicon carbide substrate surface before etching. The values in parentheses are the etch depth.

In order to evaluate the surface smoothing effect of silicon carbide by chlorine trifluoride gas, the surface roughness is measured using the root-mean-square (RMS) roughness as shown in Figure 7 (b).Figure 7 (a) also shows the etch depth. The initial surface has an RMS roughness of 5 μm. The surface roughness decreases with increasing etch period. At 10 minutes, when 180 μm has been etched, the RMS roughness becomes a low value of 1 μm. Consistent with Figures 6 (d) and (e), the RMS roughness is maintained at nearly 1 μm at 30 minutes when the etch depth becomes greater than 500 μm. Solid line in Figure 7 (b) indicates the possibility that the chlorine trifluoride gas has a smoothing effect on the 3C-silicon carbide surface.

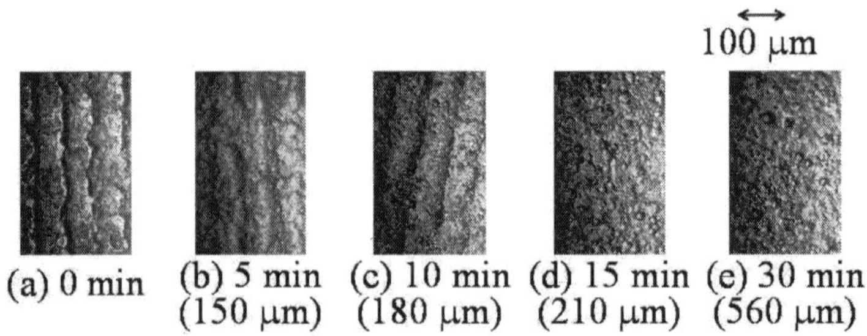

Figure 6. Photograph of the silicon carbide surface etched using chlorine trifluoride at atmospheric pressure and 770 K at (a) 0 min, (b) 5 min, (c) 10 min, (d) 15 min, and (e) 30 min. The values in parenthes are the etch depth.

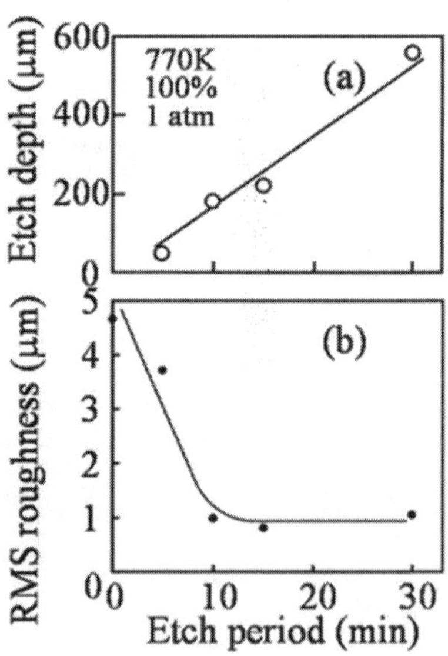

Figure 7. Etch depth and roughness of the silicon carbide surface etched by chlorine trifluoride gas at atmospheric pressure and 770 K within 30 min: (a) etch depth, and (b) root mean square (RMS) roughness.

Figure 8 shows photographs of the polycrystalline 3C-silicon carbide surface etched using a gas mixture of chlorine trifluoride and nitrogen at 670-870K for 15 min. The concentration and the flow rate of the gas mixture are 10-100% and 0.2 slm, respectively. Figure 9 shows the RMS roughness of the polycrystalline 3C-silicon carbide surface etched using chlorine trifluoride gas at atmospheric pressure and 670-870K for 15 min. The triangle, diamond and circle show the RMS roughness at the chlorine trifluoride gas concentrations of 10, 50 and 100 %, respectively.

The photograph indicated by 'Before etch' in Figure 8 shows the initial surface, which has very narrow, vague and shallow trenches formed by mechanical polishing. Using the chlorine trifluoride gas concentration of 10%, the change in the surface morphology is explained. The surface etched at 730K and 10% is recognized to have circular-shaped pits. Although the etching rate under this condition is very low, its surface shown in Figure 8 has an etched depth of 15 μm. This surface shows many circle-like pits, the edge of which is clearly shown. These may be the grain boundary or some disordered region which can be etched at a slightly higher etching rate.

At 770 K and 10 %, the shape of the circular-shaped pits still clearly exists, similar to that at 730 K and 10 %. The photograph at 870 K and 10% shows pits smaller than those at 770 K and 10%. Simultaneously, the conical shape of the pits still exists. This shows that chlorine trifluoride gas at the low concentration of 10% has a quite small role of smoothing the surface, but rather tends to roughen it. This trend is measured using the RMS roughness, shown by the triangles in Figure 9. The RMS roughness before etching is nearly 0.5 μm; it increases with the increasing substrate temperature at the chlorine trifluoride gas concentration of 10%.

Next, at the fixed chlorine trifluoride gas concentration of 50%, the change in the etched surface morphology is explained using Figure 8. The surface etched at 730 K and 50% shows the clear edge shape of the pits. The surface etched at 770 K and 50% still has a clear edge of the conical-shaped pits. Although some pits still have such the clearly observed edge shape, the rest of the surface has no clear edges. The surface morphology at 870 K and 50% shows both clear and vague edges. Since semi-smoothed and clear pits coexist there, the RMS roughness, indicated by the diamonds in Figure 9, still slightly increases with the increasing substrate temperature.

The change in the surface morphology etched at 100% is also shown in Figure 8. The surface etched at 670 K and 100% shows the clear edge of the pits. In contrast to this, the surface etched at 730 K and 100% shows the slightly vague edge of the pits. For the surface etched at 770 K and 100 %, the conical-shaped pits still remain, but are few. The edge of the conical-shaped pits disappears, when the surface is etched at 870 K. Since this trend in smoothing the surface appears in the RMS roughness behavior, the RMS roughness at 100%, indicated using circles in Figure 9, slightly decreases with the increasing substrate temperature.

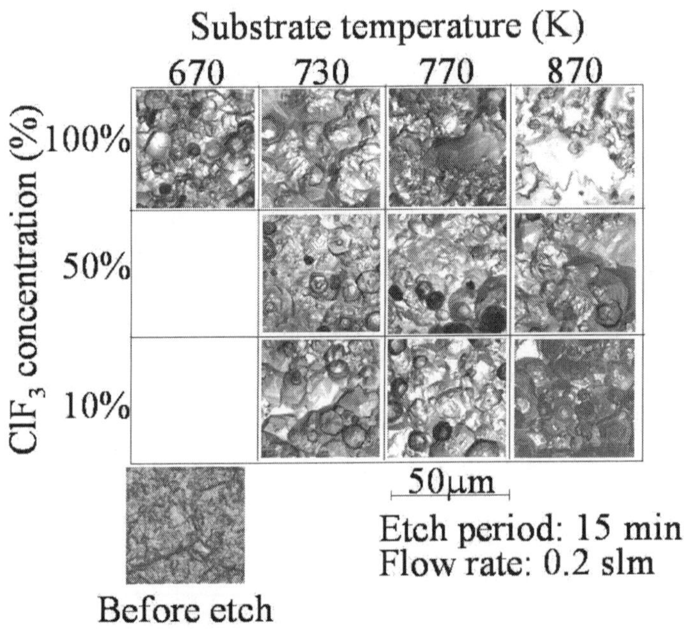

Figure 8. Photograph of the polycrystalline 3C-silicon carbide surface etched using chlorine trifluoride gas at atmospheric pressure for 15 min at 670-870 K, 10-100% and 0.2 slm.

From the view point of the effect of the etchant gas concentration at each temperature, the shape of the pits tends to become unclear with the increasing chlorine trifluoride gas concentration. Therefore, as the overall trend, the higher temperature and the higher chlorine trifluoride gas concentration produces a smoother surface of the polycrystalline 3C-silicon carbide.

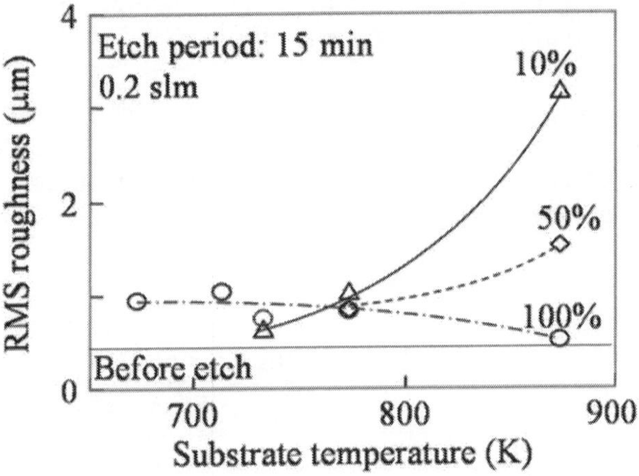

Figure 9. RMS roughness of the polycrystalline 3C-silicon carbide substrate surface etched using chlorine trifluoride gas at atmospheric pressure and 670-870 K for 15 min. Triangle: 10 %, diamond: 50 %, and circle: 100%.

The polycrystalline 3C-silicon carbide etching rate can be adjusted using the combination of gas flow rate, gas concentration and the substrate temperature, in order to obtain surfaces suitable for various purposes. This technique is expected to be used for various applications, such as the dry cleaning of the silicon carbide substrate surface instead of wet method, and the removal of the damaged layer formed during the chemical mechanical polishing using diamond slurry.

2.4. Surface Chemical Condition and Etching Rate

The fraction of silicon and carbon on the silicon carbide surface remaining after the etching is useful information for developing various processes which are performed after the etching. Thus, the chemical bonds at the silicon carbide surface are measured using X-ray photoelectron spectroscopy (XPS) before and after etching by chlorine trifluoride gas. Figure 10 shows the fraction of carbon, silicon, oxygen, chlorine and fluorine on the silicon carbide substrate before and after etching the depth greater than 150 μm using the chlorine trifluoride gas for 15 min at atmospheric pressure and at 720 K.

The silicon carbide surface before the etching has a carbon fraction nearly equal to that of silicon. However, the fraction of carbon significantly increases to 75 % after the etching. This indicates that the production of volatile carbon compound is slower than that of silicon compound at this temperature.

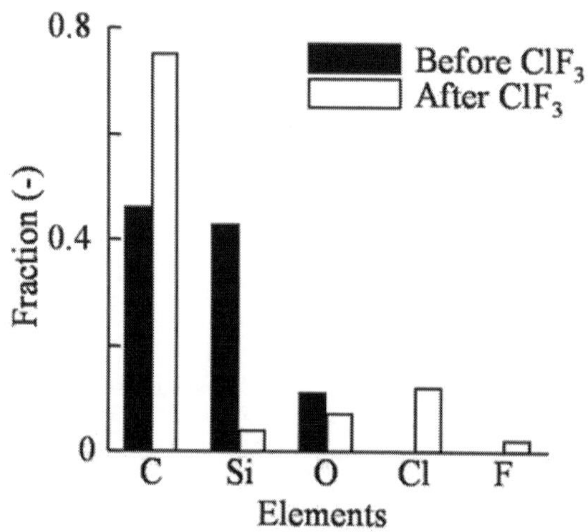

Figure 10. Fraction of carbon, silicon, oxygen, chlorine and fluorine on the silicon carbide substrate surface before and after etching using chlorine trifluoride for 15 min at atmospheric pressure in the reactor. The temperature of the silicon carbide substrate is 720 K.

In order to evaluate the state of silicon and carbon, the XPS spectra of Si 2p and C 1s are shown inFigures 11 and 12, respectively. The conditions of etching are the same as in Figure 10.

An almost amount of silicon at the silicon carbide surface has chemical bond with carbon before etching as shown in Figure 11 (a). However, after the etching, a significant amount of silicon oxides and oxidized or halogenated silicon carbide are present as shown in Figure 11 (b). The chemical bonds of carbon simultaneously change, as same as those of silicon. Figure 12 (a) shows that chemical bonds of carbon with silicon dominate at the substrate surface before etching. After etching, an almost amount of carbon has chemical bonds with carbon as shown in Figure 12 (b). The surface after etching is covered with large amount of carbon having carbon-carbon bonds. This result is consistent with the dark appearance of the silicon carbide surface after the etching.

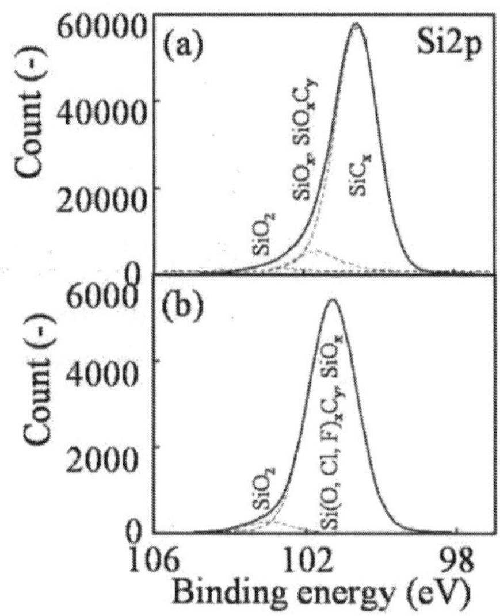

Figure 11. XPS spectra of Si 2p measured (a) before and (b) after etching using chlorine trifluoride gas for 15 min at atmospheric pressure. The temperature of the silicon carbide substrate is 720 K.

2.5. Chemical Reactions

The gaseous products and the chemical reactions associated with silicon carbide etching are explained.Figure 13 shows the mass spectra of the gaseous species existing in the exhaust gas immediately after beginning silicon carbide substrate etching at 770 K using chlorine trifluoride gas at atmospheric pressure. The partial pressures are normalized using the pressure at the mass of 28 a.m.u., which is the largest partial pressure in this measurement and which can be assigned to silicon from the silicon tetrafluoride and nitrogen remaining in the QMS system.

Typical mass spectra, shown in Figure 13, are interpreted by taking into account the fragmentation in the QMS analyzer and the isotopic abundance of chlorine [31-33]. In this figure, the ion species at a mass of 14 a.m.u. is assigned to N^+, which is the fragment of nitrogen gas. The low partial pressures corresponding to chlorine trifluoride and its fragment, $ClF_3{}^+$, at masses of 92 and 94 a.m.u., and ClF^+at 54 and 56 a.m.u., are detected. The partial pressures observed at masses of 70, 72 and 74 a.m.u. correspond to $Cl_2{}^+$, which is assigned to chlorine gas. Chlorine gas is produced due to the chemical reaction during silicon carbide etching, similar to that for the silicon etching [22, 34]. The partial pressures at masses of 35 and 37 a.m.u. can be assigned to Cl^+, which is a fragment from both chlorine trifluoride and chlorine (Cl_2). The partial pressures observed at masses of 19 and 20 a.m.u. correspond to F^+ and HF^+, respectively. F^+ and HF^+ are produced due to the fragmentation of chlorine trifluoride. Since the partial pressure of fluorine (F_2) at mass 38 a.m.u. did not appear in the mass spectra, the thermal dissociation of chlorine trifluoride gas [35] is negligible in the gas phase of the cold wall reactor used in this study.

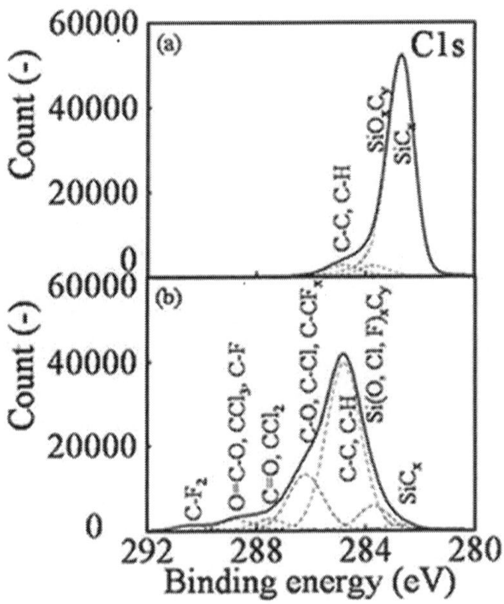

Figure 12. XPS spectra of C 1s measured (a) before and (b) after the etching using chlorine trifluoride gas for 15 min at atmospheric pressure. The temperature of the silicon carbide substrate is 720 K.

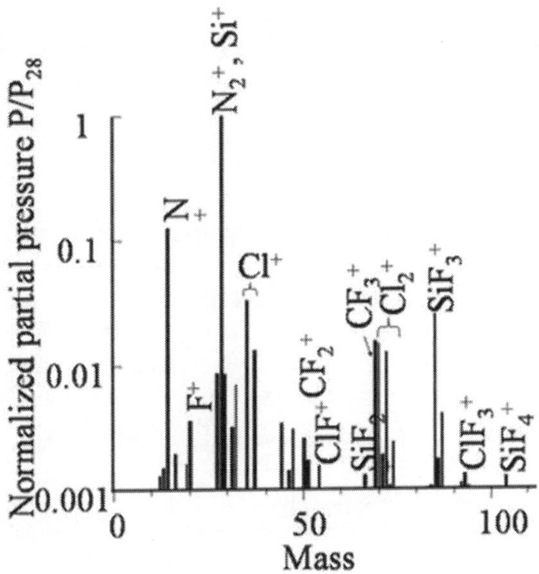

Figure 13. Mass spectra of gaseous species existing in the exhaust gas from the reactor during the etching of the silicon carbide surface using chlorine trifluoride gas at atmospheric pressure. The temperature of the silicon carbide substrate is 770 K. The ionization conditions are 70 eV and 1.73 mA.

The partial pressures at masses of 66, 85 and 104 a.m.u. can be assigned to SiF_2^{+}, SiF_3^{+} and SiF_4^{+}, respectively, whose parent is silicon tetrafluoride, like silicon etching [34]. The gaseous carbon compound produced by etching is identified as carbon tetrafluoride (CF_4), because the partial pressures at masses of 50 and 69 a.m.u. correspond to CF_2^{+} and CF_3^{+}, respectively, which can be assigned as fragments of carbon tetrafluoride.

Thus, silicon and carbon become the gaseous species of silicon tetrafluoride and carbon tetrafluoride, respectively. The overall chemical reaction between silicon carbide and chlorine trifluoride is as follows:

$$3SiC + 8ClF_3 \rightarrow 3SiF_4 + 3CF_4 + 4Cl_2 \quad (1)$$

3. SINGLE-CRYSTALLINE 4H-SILICON CARBIDE ETCHING USING CHLORINE TRIFLUORIDE GAS

3.1. Substrate, Reactor and Process

The substrate used is the n-type single-crystalline 4H-silicon carbide wafer having a (0001) surface, 8-degrees off-oriented to <11-20>. This substrate has nitrogen as the n-type dopant at the concentration of 3 - 5 x 10^{18} cm^{-3}. The 4H-silicon carbide substrate, having 5 mm wide x 5 mm long x 400 μm thick dimensions, is placed on the center of the polycrystalline 3C-silicon carbide susceptor, which has the dimension of 30 mm wide × 40 mm long × 0.2 mm thick produced by the chemical vapor deposition method (Admap Inc., Tokyo).

The reactor shown in Figure 1 is used following the process shown in Figure 2 except of hydrogen baking. The Si-face (0001) and C-face (000-1) of the 4H-silicon carbide substrates are etched using chlorine trifluoride gas. The etching is performed at the temperatures between 570 – 1570 K at the chlorine trifluoride gas flow rate of 0.1 – 0.3 slm. The average etching rate of 4H-silicon carbide is determined by the decrease in the substrate weight. Because the 3C-silicon carbide susceptor and the 4H-silicon carbide substrate are simultaneously etched in the reactor, the etching rate obtained is comparable to the average value for the wide 4H-silicon carbide substrate.

The X-ray topograph of Si-face and C-face of 4H-silicon carbide was taken at the beam-line BL15C of the Photon Factory of the High Energy Accelerator Research Organization (Proposal No. 2006G286), in order to evaluate the crystalline defects. The density and behavior of etch pit produced on the 4H-silicon carbide substrate using chlorine trifluoride gas at various temperatures were evaluated.

3.2. Numerical Calculation Of Etching Rate

The etching rate of single-crystalline 4H-silicon carbide is numerically calculated [29]. The geometry of horizontal cold-wall reactor, shown in the previous section (Figure 1), is taken into account for a series of calculations. In order to evaluate the silicon carbide etching rate and the overall rate constant in steady state in non-uniformly distributed temperature and gas flow fields, the two-dimensional equations of mass, momentum, energy, species transport and surface chemical reaction, linked with the ideal gas law, are solved. The discretized equations are coupled and solved using the SIMPLE algorithm [36] on a CFD software package, Fluent version 6 (Fluent, Inc., Lebanon, NH, USA).

The silicon carbide etching is assumed to follow the overall reaction in Eq. (2) [23, 24],

$$3SiC + 8ClF_3 \rightarrow 3SiF_4 + 3CF_4 + 4Cl_2 \qquad (2)$$

Mass changes due to the chemical reaction of Eq. (2) are taken into account in the boundary conditions at the surface of silicon carbide. The overall reaction shown in Eq. (2) is assumed to be a first-order reaction.

$$\text{SiC etching rate} = 6 \times 10^7 M_{SiC} k_{SiC} [ClF_3] / \rho_{SiC} \quad (\mu\text{mm min}^{-1}), \qquad (3)$$

where ρ_{SiC} is the density of solid silicon carbide (kg m^{-3}). M_{SiC} is the molecular weight of silicon carbide (kg mol^{-1}), respectively. The factor 6 x 10^7 is used for the unit conversion of m s^{-1} to µm min^{-1}.k_{SiC} is the overall rate constant for the reaction of Eq. (3). [ClF_3] is the concentration of chlorine trifluoride gas at the silicon carbide surface (mol m^{-3}). The concentration of each species at the surface is governed by a balance between the consumption due to the chemical reaction and the diffusion fluxes driven by the concentration.

The gas velocity and pressure at the inlet are 0.08 m s^{-1} and 1.0133 x 10^5 Pa, respectively. The heat capacities of chlorine trifluoride, nitrogen, tetrafluorosilane, tetrafluorocarbon and chlorine are taken from the literature [37]. The gas properties, such as the viscosity and the thermal conductivity of chlorine trifluoride, tetrafluorosilane, chlorine and nitrogen are estimated with the method described in the literature [38]. The Lennard-Jones parameters of σ and ε/k for chlorine trifluoride are 4.63 Angstroms and 355 K, respectively, which are obtained using a theoretical equation [38] taking the value of viscosity [39] into account. Each physical constant is expressed as a function of temperature. The properties of the mixed gas are estimated theoretically [40]. The binary diffusion coefficients of chlorine trifluoride, tetrafluorosilane and chlorine are estimated using the method described in the literature [38].

The overall rate constant, k_{SiC} in Eq. (3) is obtained so that the calculated etching rates agree with those measured at various conditions.

3.3. Etching Rate

Figure 14 shows the etching rate of the Si-face and C-face of 4H-silicon carbide at the substrate temperatures between 570 K and 1570 K. The etching rate of the C-face of 4H-silicon carbide is slightly higher than that of the Si-face of 4H-silicon carbide. The etching rate of the Si-face and C-face of 4H-silicon carbide is near 5 μm min^{-1} and it is still flat at the temperatures between 770 K and 1570 K. This flat etching rate behavior is similar to that of polycrystalline 3C-silicon carbide, shown inFigure 3 in the previous section. The various surface morphologies at higher temperatures shown in the latter part of this section are obtained at nearly the same etching rate.

Figure 15 shows the typical behavior of the etching rate changing with the chlorine trifluoride gas concentration, obtained at 1370 K. The etching rate is proportional to the chlorine trifluoride gas concentration, similar to that of polycrystalline 3C-silicon carbide, shown in Figure 4. The fluctuation of the etching rate, shown in Figures 14 and 15, is entirely 20 %, which is due to the considerable distribution of the etching rate over the 3C-silicon carbide susceptor.

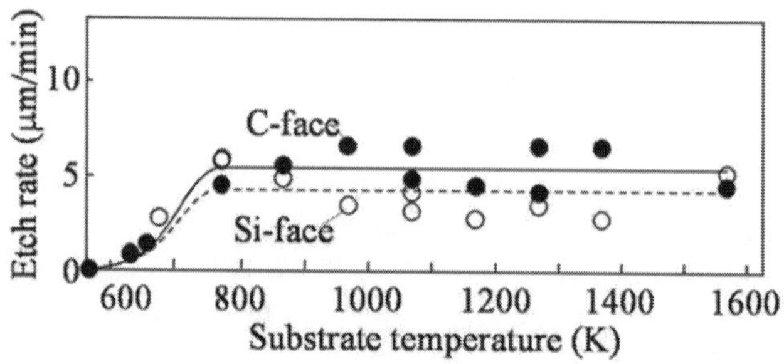

Figure 14. Etching rate of 4H-silicon carbide using chlorine trifluoride gas at 100 %, 0.1 slm and various temperatures. Dark circle: C-face, and white circle: Si-face.

The relationship between the etching rate and the chlorine trifluoride gas flow rate is shown in Figures 16. Figures 16 (a) and 16 (b) show the etching rate of the Si-face and the C-face, respectively, of the 4H-silicon carbide substrate by chlorine trifluoride gas (100%) at atmospheric pressure, when changing with the chlorine trifluoride gas flow rate. The circle, square and triangle denote the etching rate at the substrate temperatures of 770, 870 and 970 K, respectively. As shown in these figures, the Si-face and C-face etching rates increase with the gas flow rate of the chlorine trifluoride gas. Additionally, the etching rate at 770, 870 and 970 K overlap each other for both the Si-face and the C-face, consistent with Figures 14. The etching rate of the Si-face is about 60 % of that of the C-face. The relationship between the Si-face etching rate and the C-face etching rate is similar to that of another empirically known etching technique, such as the potassium hydroxide method [41].

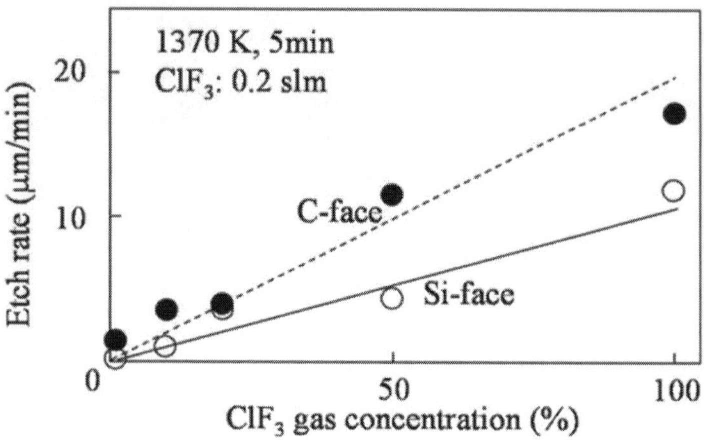

Figure 15. Etching rate of Si-face (white circle) and C-face (dark circle) 4H-silicon carbide changing with chlorine trifluoride gas concentration, at 1370 K and at the total gas flow rate of 0.2 slm for 5 min.

3.4. Surface Reaction Rate Constant

Figure 17 is the Arrhenius plot of the rate constants for etching of Si-face and C-face of 4H-silicon carbide. The rate constants is expressed in Eqs. (4) and (5).

$$k_{SiC} = 4.1 \exp(-6.6 \times 10^4 / RT) \quad (m/s) \quad \text{for Si-face} \tag{4}$$

and,

$$k_{SiC} = 98 \times \exp(-8.3 \times 10^4 / RT) \quad (m/s) \qquad \text{for C-face} \tag{5}$$

where R is the gas constant (J mol^{-1} K^{-1}).

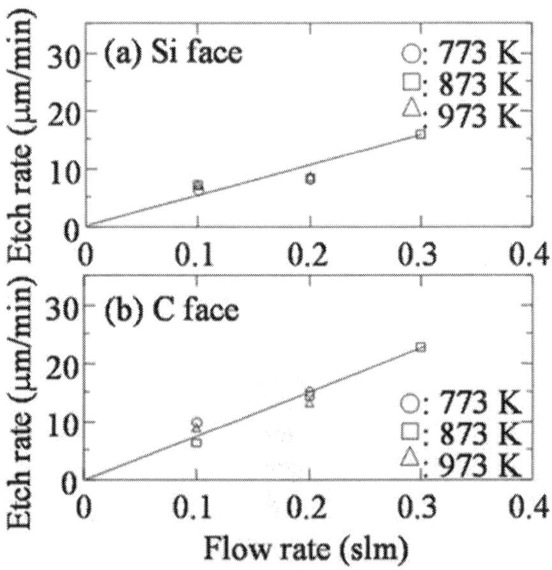

Figure 16. Etching rate of (a) Si-face and (b) C-face of 4H-silicon carbide substrate by chlorine trifluoride gas (100%) at atmospheric pressure in the flow rate range between 0.1 and 0.3 slm. Circle, square and triangle show the etching rates at the substrate temperatures of 770, 870 and 970 K, respectively.

In order to show that the rate constant of Eq. (4) can reproduce the measured etching rate behavior, the measured and the calculated etching rate values of Si-face are shown in Figure 18, as the Arrhenius plot. The calculation shows that the etching rate at the temperatures near 670K is near 1 μm/min, and it becomes near 10 μm/min at 1000 K. The calculated etching rate tends to become flat at the higher temperatures at the chlorine trifluoride gas flow rate of 0.1 and 0.2 slm. Additionally, the etching rate obtained by the calculation increases with increasing the chlorine trifluoride gas flow rate. Because the great etchant flow rate can moderate the etchant depletion occurring in the downstream region, the average etching rate over the 4H-silicon carbide substrate and the 3C-silicon carbide susceptor can increase with the increasing flow rate of chlorine trifluoride gas. The etching rate for C-face calculated using Eq. (5) also showed the typical behavior of the measurement. Therefore, Eqs. (4) and (5) are applicable to reproduce the behavior of the 4H-silicon carbide (Si-face and C-face) etching rate.

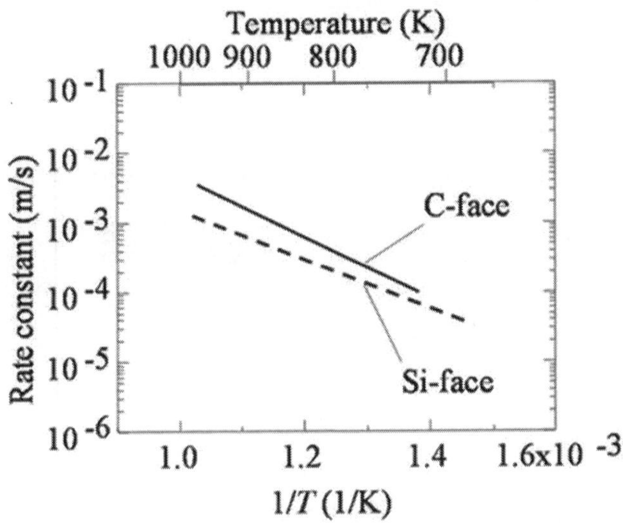

Figure 17. Rate constants for etching of Si-face (broken line) and C-face (solid line) of 4H-silicon carbide using chlorine trifluoride gas, obtained by numerical calculation.

3.5. Surface morphology

Figure 19. shows the surface morphology of the Si-face of 4H-silicon carbide before and after the etching at the chlorine trifluoride gas concentration of 100 % and at the flow rate of 0.1 slm. The etching was performed at the substrate temperatures of (b) 570, (c) 620, (d) 770, (e) 970, (f) 1270, (g) 1370 and (h) 1570 K. The etched depth was 5-18 μm.

Figure 19 (a) shows the surface morphology of the Si-face of 4H-silicon carbide substrate before the etching. Figure 19 (b) shows that there are many small pits after the etching at 570 K. At 620 K, the pits are very large, nearly 50 μm in diameter, as shown in Figure 19 (c). With the increasing temperature, the pits tends to become small and shallow, as shown in Figures 19 (d) - (h). The pit diameter after etching at 770 K, shown in Figure 19 (d), is nearly 10 % of that at 620 K, shown inFigure 19 (c). Figure 19 (e) shows that the pit diameter becomes even smaller and the pit density decreases at 970 K. This trend is very clear at the temperatures higher than 1270 K, as shown inFigures 19 (f), (g) and (h). Figure 19 (g) shows that the pit density significantly decreases at 1370 K. The sharp-shaped pits, presented at the lower temperatures, are not there, on the silicon carbide surface after the etching at 1570 K, as shown in Figure 19 (h).

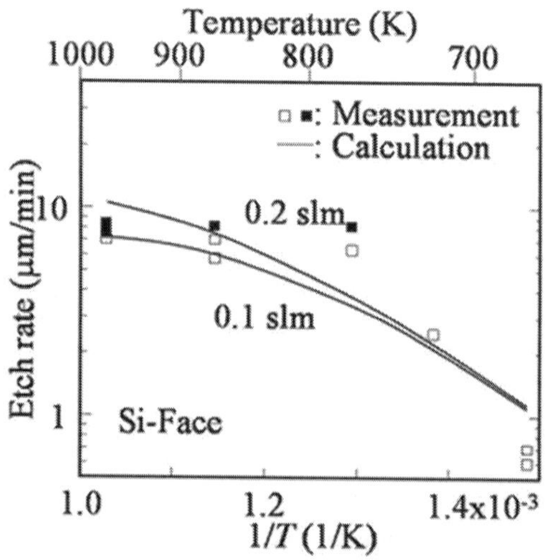

Figure 18. Arrhenius plot of 4H-silicon carbide Si-face etching rate at chlorine trifluoride flow rate of 0.1 and 0.2 slm. Square: measurement, solid line: calculation.

Next, the influence of chlorine trifluoride gas concentration on the surface morphology is explained.Figure 20 shows the surface morphology of the Si-face 4H-silicon carbide surface taken by the optical microscope before and after the etching for 5 min at the total flow rate of 0.2 slm, at the substrate temperature of 1370 K and at various chlorine trifluoride gas concentrations. Figure 20 (a) shows the surface morphology of the Si-face of 4H-silicon carbide before the etching.

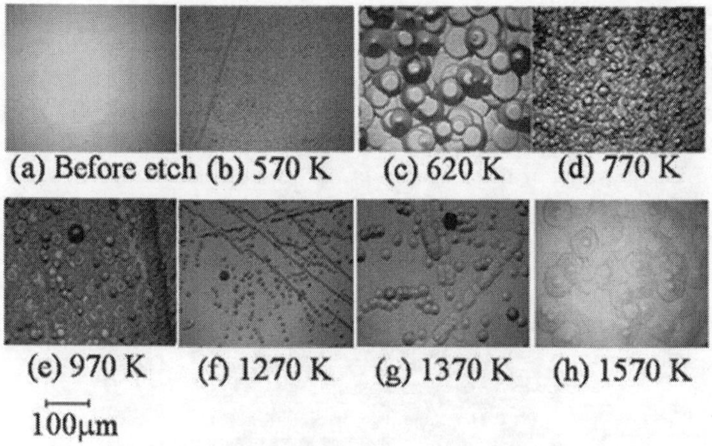

Figure 19. Surface morphology of Si-face of 4H-silicon carbide (a) before and after the etching using chlorine trifluoride gas at the concentration of 100 %, at the substrate temperature of (b) 570, (c) 620, (d) 770, (e) 970, (f) 1270, (g) 1370 and (h) 1570 K and at the flow rate of 0.1 slm. The etched depth is 5-18 μm.

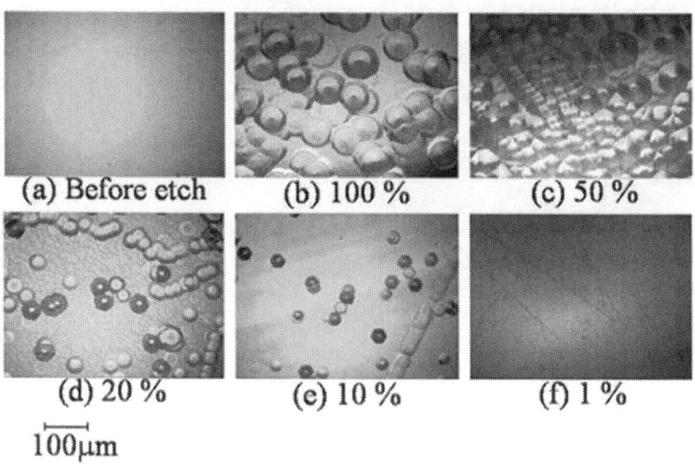

Figure 20. Surface morphology of Si-face 4H-silicon carbide surface (a) before and after etching at the total gas flow rate of 0.2 slm, 1370 K and various chlorine trifluoride gas concentrations of (b) 100, (c) 50, (d) 20, (e) 10 and (f) 1%, for 5 min.

Although there are many large pits at the chlorine trifluoride gas concentration of 100%, as shown inFigure 20 (b), they become significantly small and less at 20%, as shown in Figure 20 (d). Figure 20 (e)shows that the surface etched at 10% is flat with only a small number of pits. At 1%, most of the surface is flat, except for scratches, as shown in Figure 20 (f).

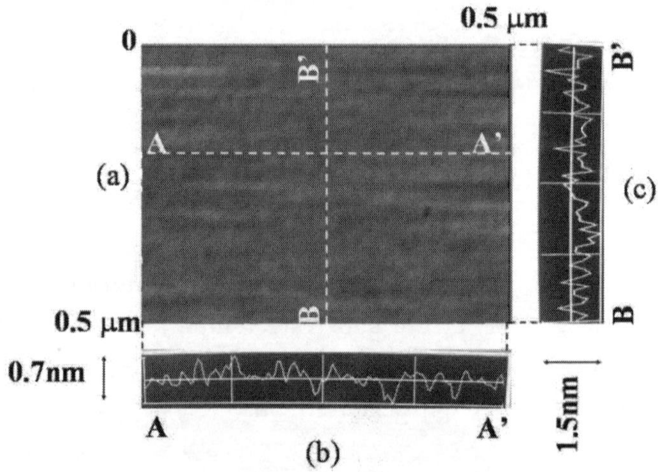

Figure 21. AFM photograph of Si-face of 4H-silicon carbide, etched for 0.5 min using the chlorine trifluoride gas concentration of 1% diluted in ambient nitrogen at the substrate temperature of 1570 K and at the total flow rate of 4 slm. (a): plan view, (b): A-A' cross section, and (c): B-B' cross section.

From Figures 19 and 20, the Si-face 4H-silicon carbide surface after etching tends to be flat with the increasing temperature and decreasing chlorine trifluoride gas concentration. Following this trend, the Si-face 4H-silicon carbide surface is etched at 1570 K at the chlorine trifluoride gas concentration of 1% for 0.5 min. Figure 21 shows the AFM photograph of the etched surface. Figures 21 (a), (b) and (c)are the plan view, A-A' cross section and B-B' cross section, respectively. Although the etched depth is only about 0.03 μm, it can reveal the trend of the surface, causing pit or not. As shown in Figure 21 (a), this surface does not show any etch pit; Figures 21 (b) and (c) shows no periodical shape reflecting the 4H-silicon carbide crystal step [42]. The root-mean-square (RMS) roughness are 0.1 and 0.2 nm on A-A' line and B-B' line, respectively, which are comparable to that of the polished 4H-silicon carbide substrate surface. Thus, the shallow etching for removing thin layer, such as damaged layer, is possible with maintaining the specular surface of the Si-face of 4H-silicon carbide.

Figure 22. shows the surface morphology of the C-face of 4H-silicon carbide before and after the etching at the chlorine trifluoride gas concentration of 100 %, at (b) 570, (c) 620, (d) 770, (e) 1070, (f) 1270, (g) 1370 and (h) 1570 K and at the flow rate of 0.1 slm. The etched depth is 10-30 μm. Figure 22 (a) is the C-face 4H-silicon carbide surface before the etching.

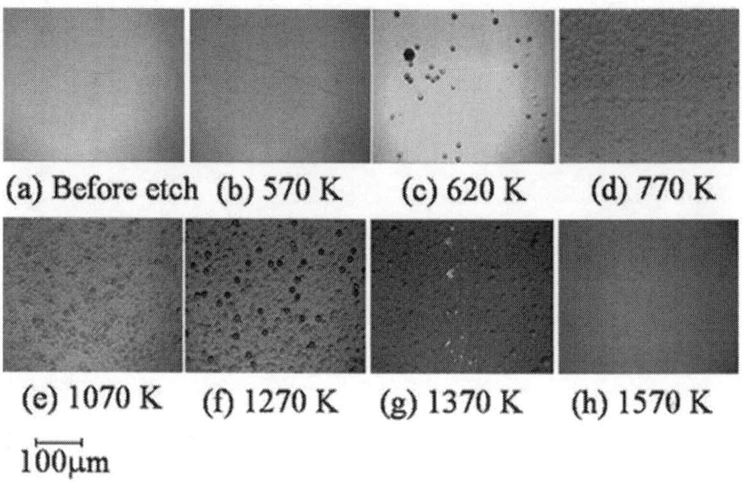

Figure 22. Surface morphology of C-face of 4H-silicon carbide (a) before and after the etching using chlorine trifluoride gas at the concentration of 100 %, at the substrate temperature of (b) 570, (c) 620, (d) 770, (e) 1070, (f) 1270, (g) 1370 and (h) 1570 K and at the flow rate of 0.1 slm. Etched depth is 10-30 μm.

There is the flat surface after the etching at 570 K, as shown in Figure 22 (b), because of significantly small etching rate. However, Figure 22 (c) shows that pits are produced at 620 K. The surface etched at the temperatures between 770 K and 1270 K have many small pits, as shown in Figures 22. (d), (e) and (f). Figures 22 (g) and (h) show that the pit diameter decreases at the temperatures higher than 1370 K. The surface etched at 1570 K shows a flat surface as shown in Figure 22 (h).

Next, the influence of the chlorine trifluoride gas concentration on the surface morphology of the C-face of 4H-silicon carbide is explained. Figure 23 shows the morphology of the C-face 4H-silicon carbide surface before and after the etching for 5 min at the various chlorine trifluoride gas concentrations of (b) 100, (c) 50, (d) 20, (e) 10 and (f) 1% at the total flow rate of 0.2 slm. The substrate temperature is fixed at 1370 K. The etched depth is 3-84 μm.

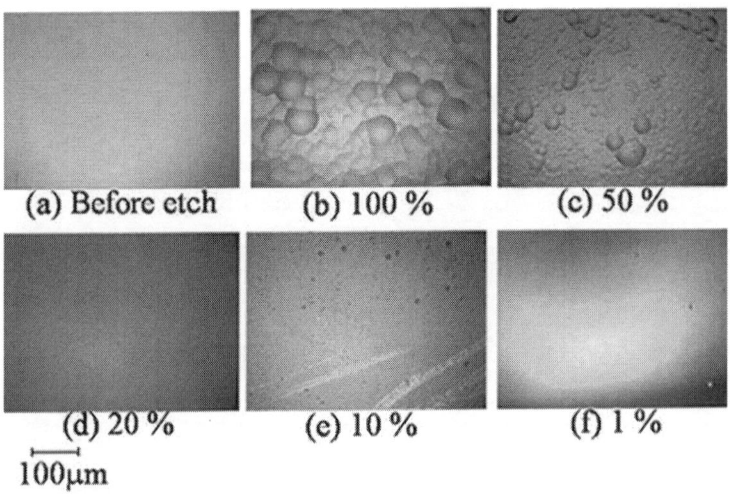

Figure 23. Surface morphology of C-face 4H-silicon carbide surface before and after etching at the total gas flow rate of 0.2 slm, 1370 K and various chlorine trifluoride gas concentrations for 5 min. (a) before etching, (b) 100, (c) 50, (d) 20, (e) 10 and (f) 1%. Etched depth is 3-84 μm.

Figure 23 (b) shows that the C-face 4H-silicon carbide surface has large and shallow pits after etching at the chlorine trifluoride gas concentration of 100%. As shown in Figure 23 (c), the etch pits become shallow at the chlorine trifluoride gas concentration of 50%. Figures 23 (d), (e) and (f) show that the etched surface is entirely flat at the chlorine trifluoride gas concentrations less than 20%. Particularly, the surface etched at 1% is flat, as shown in Figure 23 (f). Overall, the trend in the etched surface morphology of the C-face of 4H-silicon carbide is similar to that of the Si-face of 4H-silicon carbide, although the pit size of the C-face is smaller than that of Si-face.

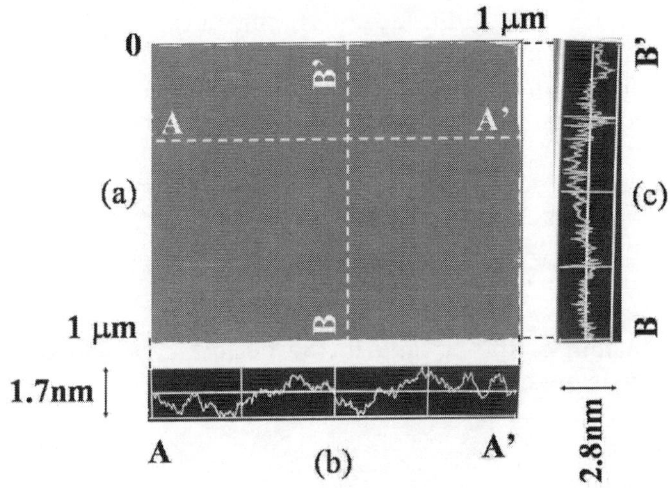

Figure 24. AFM photograph of C-face of 4H-silicon carbide, etched for 0.5 min using the chlorine trifluoride gas concentration of 1% diluted in ambient nitrogen at the substrate temperature of 1570 K and at the total flow rate of 4 slm. (a): plan view, (b): A-A' cross section, and (c): B-B' cross section.

Similar to the Si-face 4H-silicon carbide surface, the C-face 4H-silicon carbide surface is etched at 1570 K at the chlorine trifluoride gas concentration of 1% for 0.5 min. Figures 24 (a), (b) and (c) are the AFM photographs of the plan view, A-A' cross section and B-B' cross section, respectively. The etched depth is near 0.03 μm. Figure 24 (a) does not show any shape like an etch pit; Figures 24 (b) and (c)show no periodical shape reflecting the 4H-silicon carbide crystal step [42]. The RMS roughness is 0.4 and 0.3 nm on A-A' line and B-B' line, respectively, which are comparable to that of the polished 4H-silicon carbide substrate surface. Thus, the shallow etching without producing any trace of pit shape is possible, for the C-face of 4H-silicon carbide.

3.6. Surface Morphology Behavior and Its Rate Process

Entire surface morphology behavior changing with the substrate temperature for Si-face and C-face of 4H silicon carbide is summarized as (i) very small change at very low temperatures (lower than 570 K), (ii) significant pit formation between 570 K and 1270 K, and (iii) pit formation reduced at high temperatures (higher than 1370 K).

The process of pit formation is described following the rate theory, assuming that the etch pit is formed due to the difference of the etching rate between the perfect crystal region and the weak spot having any kinds of damage and crystalline defect [10].

The rate constant of the etching in the perfect crystal region, k_P, is assumed to be expressed in Eq. (6).

$$k_P = A\exp\left(-\frac{E}{RT}\right) \tag{6}$$

where A is the pre-exponential factor, E is the activation energy, R is the gas constant, and T is the substrate temperature. In contrast to this, the weak spot, which has larger etching rate to cause pit, is assumed to have the slightly smaller activation energy than that in the perfect region. The rate constant at the weak spot, k_W, is assumed to be expressed in Eq. (7), using the difference of the activation energy from that in the perfect region, ΔE.

$$k_W = A\exp\left(-\frac{E-\Delta E}{RT}\right) \tag{7}$$

Assuming that the etchant gas concentration is the same in the perfect region and at the weak spot, the pit depth is expressed in Eq. (8), taking into account that $\Delta E/RT$ is very small.

$$Pit depth = V_E\left(\frac{k_W - k_P}{k_P}\right) = V_E\left(\exp\left(\frac{\Delta E}{RT}\right) - 1\right) \cong V_E\frac{\Delta E}{RT} \tag{8}$$

where V_E is the etching rate in the perfect region.

Here, assuming that V_E shown in Figure 14 is the etching rate in the perfect region, the normalized pit depth, *h*, is evaluated and shown in Figure 25. The *h* value is defined using the maximum value of the pit depth, in Eq. (9).

$$h = \frac{Pit\,depth}{Pit\,depth_{MAX}} = \frac{\frac{V_E}{T}}{\left(\frac{V_{E,}}{T}\right)_{Pit\,depth_{MAX}}} \tag{9}$$

In Figure 25, the *h* value at the temperatures lower than 500 K is very small; it significantly increases near 700 K. After showing its maximum, the *h* value gradually decreases with the increasing substrate temperature. Near 1600 K, the *h* value is significantly smaller than the maximum value. This trend qualitatively agrees with that of the 4H-silicon carbide surface etched using chlorine trifluoride gas. Thus, the surface morphology trend over wide temperature range can be understood mainly by the rate process.

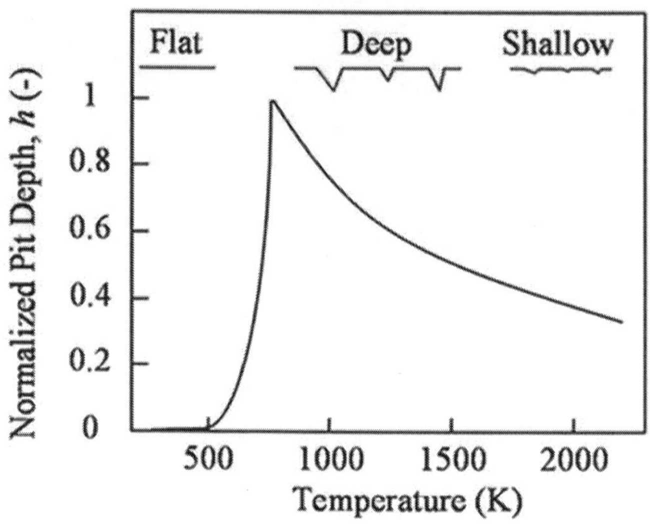

Figure 25. Normalized pit depth and temperature-dependent surface morphology behavior following the rate theory.

3.7. Etch Pits and Crystalline Defect

AFM photographs of the pit shape formed by the chlorine trifluoride gas are shown in Figure 26.Figures 26 (a) and (b) show the pits on the Si-face and C-face, respectively, after the etching using the chlorine trifluoride gas (100%) at atmospheric pressure for 3 min at 870 K and 0.3 slm. Figure 26 (c) is the etch pit of Figure 26 (a), the edge of which is traced using dotted line.

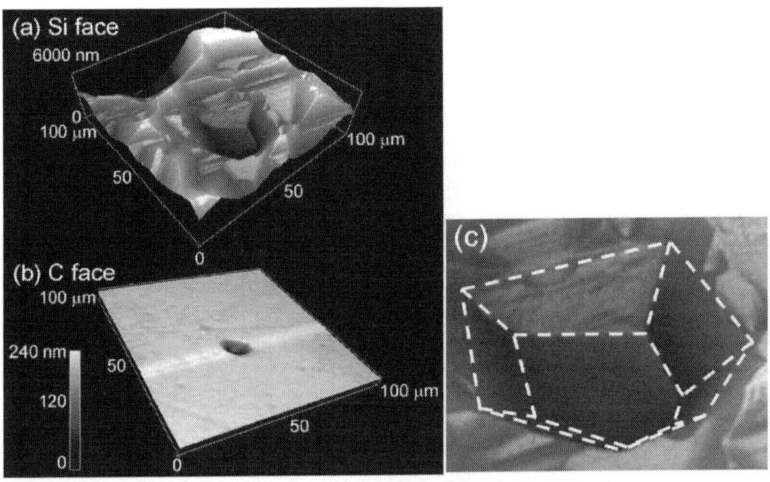

Figure 26. Etch pits on (a) Si-face and (b) C-face (870 K, chlorine trifluoride: 100 %, 0.3 slm). (c) is the magnification of the pit in (a), the edge of which is traced using dotted line.

Figures 26 (a) and (c) reveals a nearly hexagonal edge shape having a flat-shaped bottom. Its diameter and depth are about 0.03 mm and 2000-3000 nm, respectively. The many pits that exist over the entire surface are shown to have the same edge and bottom shape as that shown in Figure 26 (a). Figure 26 (b) shows the pit shape on the C-face. This surface is very flat and smooth having only a very small number of circular shaped pits, which are very shallow with a diameter of 0.01 mm and a depth of about 200 nm.

Many etch pits at the 4H-silicon carbide surface, produced by the etching using chlorine trifluoride gas, are expected to show a relationship with various crystalline defects, when the etching condition is appropriate. Thus, the Si-face and C-face 4H-silicon carbide surface are etched using chlorine trifluoride gas at 100% and at 700 K so that the pit depth become maximum value, as predicted byFigure 25. Additionally, the etch pits are compared with the X-ray topograph, because the X-ray topograph is suitable in order to evaluate an origin of etch pit [42, 43].

Figures 27 (a) and 28 (a) are the X-ray topograph of the Si-face and C-face 4H-silicon carbide surface, respectively. Figures 27 (b) and 28 (b) are the photograph of Si-face and C-face 4H-silicon carbide surface, respectively, etched using the chlorine trifluoride gas at 100% and at 700 K for 60 min. White arrows in these figures indicate the position of the spots in the X-ray topograph and the etch pits at the etched surface.

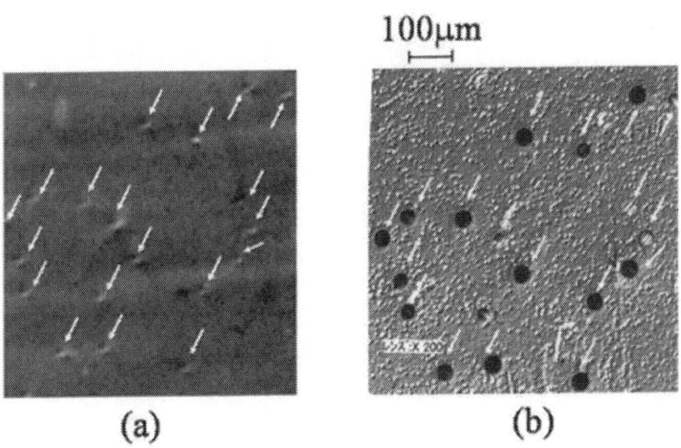

Figure 27. Comparison between the X-ray topograph and the etched Si-face 4H-silicon carbide surface. (a) X-ray topograph of the Si-face 4H silicon carbide surface, and (b) the Si-face 4H-silicon carbide surface etched using chlorine trifluoride gas at 100% and at 700 K for 60 min. Arrows in this figure indicate the spot in the X-ray topograph and the etch pit at the etched surface.

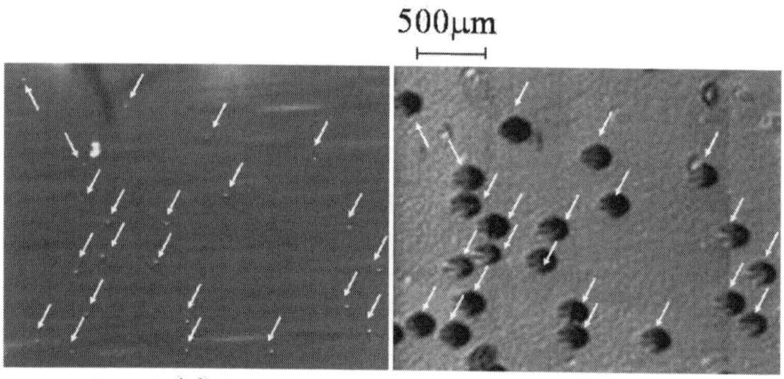

Figure 28. Comparison between the X-ray topograph and the etched C-face 4H-silicon carbide surface. (a) X-ray topograph of the Si-face 4H silicon carbide surface, and (b) the C-face 4H-silicon carbide surface etched using chlorine trifluoride gas at 100% and at 700 K for 60 min. Arrows in this figure indicate the spot in the X-ray topograph and the etch pit at the etched surface.

As indicated using white arrows, there are many etch pits, the positions of which correspond to those of the spots in the X-ray topograph. The dimension of spot in Figure 28 (a) is larger than that in Figure 27 (a); the diameter of etch pits in Figure 28 (b) is about 250 μm which is similarly larger than that inFigure 27 (b), about 40 μm. Taking into account the report [44] about the dimension of etch pits formed by KOH, Figures 28 (a) and (b) may show the screw dislocation; Figures 27 (a) and (b) may show the threading edge dislocation. Because the etching technique using the chlorine trifluoride gas may reveal the crystalline defects, like the KOH technique [44], the relationship between etch pits and various crystalline defects should be further studied.

Because of the functions to produce the specular surface and to reveal the crystalline defects, chlorine trifluoride gas is expected to be more useful than the other wet and dry techniques [41], for silicon carbide industrial process.

Next, the density and behavior of etch pit produced on the C-face of 4H-silicon carbide substrate using chlorine trifluoride gas at various temperatures were evaluated. The etch pit obtained using the chlorine trifluoride gas at 713 K for 10 min at 100% was observed and shown in Figure 29. The diameter of etch pits was near 10 μm, which was considered to be assigned to screw dislocations, following the previous study [28].

Because the etch pit density (EPD) changed with the substrate temperature [28], the etching was performed at various temperatures around 713 K. The etch pit density obtained in an area of 500 x 500 μm^2 at various substrate temperatures is shown in Figure 30. At the temperatures below 673 K, the etch pit density was very small. At 683 K, the etch pit density increase to the value near 2 x 10^4 cm^{-2}. At 713K, the etch pit density showed the maximum value of 4 x 10^4 cm^{-2}. Although the etch pit density decreased at the temperatures higher than 723 K, its value maintained near 10^4 cm^{-2}.

The etch pit density obtained at 713 K coincided with the values widely accepted as the current dislocation density level of 4H silicon carbide. Thus, the etch pit density of C-face of 4H-silicon carbide obtained in this study is acceptable and is expected to show a relationship with the crystal quality. The etch pits obtained in this study were classified to the large circular-shaped and small oval-shaped pits, which ratio were 90% and 10%, respectively, at the etching temperature of 713 K. The former is considered to be assigned to the threading screw dislocation, and the latter can be the threading edge dislocation.

In order to show that the origin of these etch pit was assigned to the crystalline imperfection, the etched surface was further etched using the same etching condition. Figure 31 (a) shows the nine etch pits observed on the C-face of 4H silicon carbide after the etching at 713 K for 10 min using the 100 % chlorine trifluoride gas. Their diameter was about 10 – 15 μm. This substrate was additionally etched at 713 K at the 100 % chlorine trifluoride gas. Figure 31 (b) shows the surface, 80 μm of which surface was etched off by the additional etching. Figure 31 (b) shows the large and shallow nine etch pits, which were overlapped with each other. The number and the position of etch pit center in Figure. 31 (b) corresponded to those in Figure 31 (a), respectively. Thus, over the depth of 80 μm, the origin to cause the etch pit was concluded to exist; the origin could be the dislocation, such as the threading dislocation, existing normal to the substrate surface.

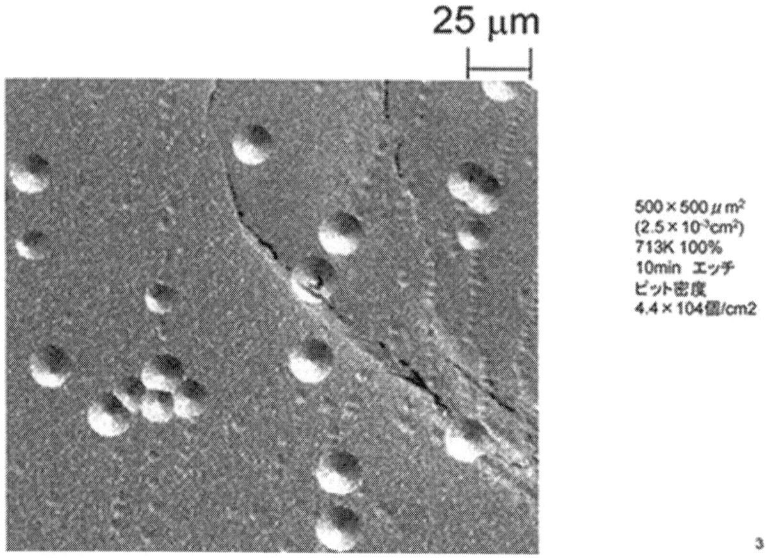

Figure 29. Etch pits produced on C-face 4H-silicon carbide surface by 100% chlorine trifluoride gas at 713 K for 10 min

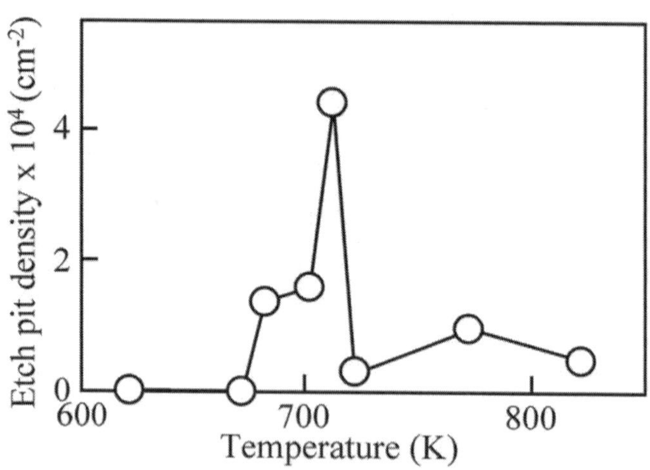

Figure 30. Etch pit density on C-face 4H-silicon carbide surface produced by chlorine trifluoride gas at various substrate temperatures.

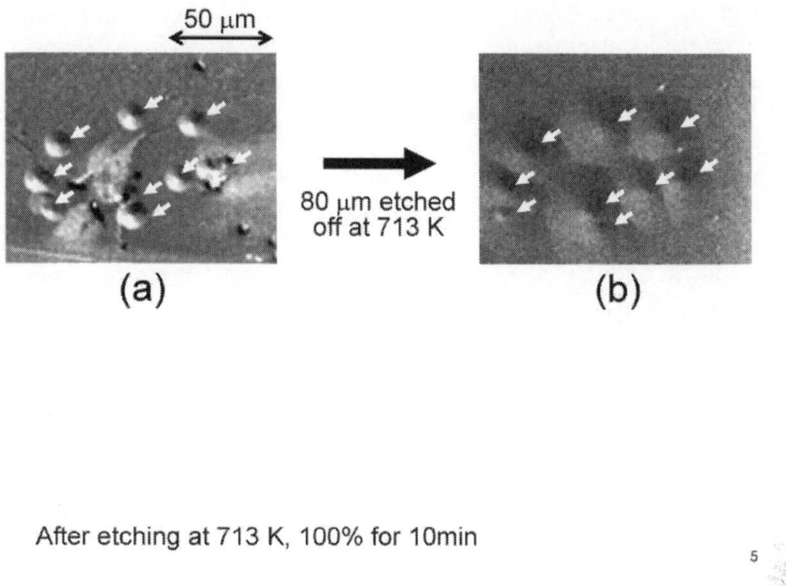

Figure 31. Comparison of etch pits (a) before and (b) after additional etching. White arrows indicate the etch pits.

The results obtained in this study indicate that the pits caused by the etching using chlorine trifluoride gas at around 713 K has the origin of crystalline imperfection, such as the threading dislocation. However, the interpretation of etch pits with respect to the underlying dislocation type should be further carefully performed by a comparative investigation [28]. Additionally, the density and shape of etch pit by this technique should be further clarified and verified through many characterization.

4. SUMMARY

Silicon carbide etching using chlorine trifluoride gas with high etching rate occurs at the temperatures higher than 770 K. Its chemical reaction is as follows:

$$3SiC + 8ClF_3 \rightarrow 3SiF_4 + 3CF_4 + 4Cl_2$$

The etching rate is 10-20 μm/min and 5 μm/min, for polycrystalline 3C-silicon carbide and single-crystalline 4H-silicon carbide, respectively. The etching rate of Si-face of 4H-silicon carbide is slightly smaller than that of C-face. The etched surface tends to be carbon rich. The etched surface of Si-face of 4H-silicon carbide shows various kinds of morphology: pitted at low temperatures of 570 – 1270 K, and smooth at 1570 K. The C-face of 4H-silicon carbide shows the similar trend, and is entirely very smoother than that of Si-face. Most of the etch pits formed near 700 K at the Si-face and C-face of 4H-silicon carbide show relationship between dislocations revealed by the X-ray topograph. At the substrate temperature of 713 K, the etch pit density showed the maximum value of 4×10^4 cm^{-2}. The etch pit density obtained by this technique is considered to show the crystal quality, particularly, the dislocation density.

ACKNOWLEDGEMENTS

Various studies reviewed in this Chapter were performed with Ms. Satoko Oda, Mr. Yusuke Katsumi, Mr. Yu Kasahara, Ms. Keiko Tanaka, Mr. Kazuchika Furukawa, Mr. Yusuke Fukumoto, Dr. Yutaka Miura, Mr. Yoichi Negishi, Dr. Takashi Takeuchi, Prof. Masahiko Aihara, Prof. Minoru Takeda, Prof. Hironobu Kunieda and Prof. Kenji Aramaki of Yokohama National University, Mr. Yasushi Fukai, Mr. Katsuya Fukae, Mr. Naoto Takechi, Dr. Yuan Gao, and Mr. Shinji Iizuka of Kanto Denka Kogyo Co., Ltd., and Dr. Tomohisa Kato, Dr. Hajime Okumura and Dr. Kazuo Arai of National Institutes of Advanced Science and Technology. Mr. N. Okumura of Keyence Co., Ltd., is very much appreciated for the surface roughness evaluation. X-ray topography experiment has been performed under the approval of the Photon Factory Program Advisory Committee (Proposal No. 2006G286).

REFERENCES

1. Y. Gogotsi, S. Welz, D. A. Ersoy, M. J. Mc Nallan, 2001 Nature411 6835 283

2. K. Vyshnyakova, G. Yushin, L. Pereselentseva, Y. Gogotsi, 2006 Int. J. Appl. Ceramic Tech.3 485

3. A. V. Zinovev, J. F. Moore, J. Hryn, M. J. Pellin, 2006 Surf. Sci600 2242

4. M. Cooke, 2005 III-Vs Review18 Dec40

5. C. Chai, Y. Yang, Y. Li, H. Jia, 2003 Optical Materials23 103

6. M. Mehregany, C. A. Zorman, S. Roy, A. J. Fleischman, C. H. Wu, N. Rajan, 2000 International Materials Reviews45 85

7. C. R. Stoldt, C. Carraro, W. R. Ashurst, D. Gao, R. T. Howe, R. Maboudian, 2002 Sensors and Actuators a-Physical97-98 410

8. N. Rajan, M. Mehregany, C. A. Zorman, S. Stefanescu, T. P. Kicher, 1999 Journal of Microelectromechanical Systems8 251

9. W. R. Ashurst, M. B. J. Wijesundara, C. Carraro, R. Maboudian, 2004 Tribology Letters17195

10. F. Shimura, 1989 Semiconductor Silicon Crystal Technology244 Academic Press San Diego, USA

11. B. Kim, S. Kim, S. Ann, B. Lee, 2003 Thin Solid Film434 276

12. U. Schmid, M. Eickoff, C. Richter, G. Kroetz, D. Schmitt-Landsiedel, 2001 Sensors and Actuators A9487

13. R. Wolf, R. Helbig, 1996 J. Electrochem. Soc.143 1037

14. P. H. Yih, A. J. Steckl, 1995 J. Electrochem. Soc., 142 312.

15. Z. Y. Xie, C. H. Wei, L. Y. Li, Q. M. Yu, J. H. Edgar, 2000 J. Cryst. Growth217 115

16. E. K. Sanchez, S. Ha, J. Grim, M. Skowronski, W. M. Vetter, M. Dudley, R. Bertke, W. C. Mitchel, 2002 J. Electrochem. Soc149 G 131

17. M. Syvajarvi, R. Yakimova, E. Janzen, 2000 J. Electrochem. Soc147 3519

18. J. S. Shor, X. G. Zhang, R. M. Osgood, 1992 J. Electrochem. Soc139 1213

19. P. Chabert, 2001 J. Vac. Sci. Technol. B, 19 1339

20. P. A. Khan, B. Roof, L. Zhou, I. Asesida, 2001 J. Electron. Mater.30 212

21. L. Jiang, R. Cheung, R. Brown, A. Mount, 2003 J. Appl. Phys.93 1376

22. H. Habuka, H. Koda, D. Saito, T. Suzuki, A. Nakamura, T. Takeuchi, M. Aihara, 2003 J. Electrochem. Soc.150 G461 464

23. H. Habuka, S. Oda, Y. Fukai, K. Fukae, T. Sekiguchi, T. Takeuchi, M. Aihara, 2005 Jpn. J. Appl. Phys44 1376

24. H. Habuka, S. Oda, Y. Fukai, K. Fukae, T. Takeuchi, M. Aihara, 2006 Thin Solid Films514 193

25. Y. Miura, Y. Katsumi, S. Oda, H. Habuka, Y. Fukai, K. Fukae, T. Kato, H. Okumura, K. Arai, 2007 Jpn. J. Appl. Phys, 46 7875

26. H. Habuka, Y. Katsumi, Y. Miura, K. Tanaka, Y. Fukai, T. Fukae, Y. Gao, T. Kato, H. Okumura, K. Arai, 2008 Materials Science Forum, 600-603 655

27. H. Habuka, K. Tanaka, Y. Katsumi, N. Takechi, K. Fukae, T. Kato, 2010 Materials Science Forum 645-648 787

28. H. Habuka, K. Tanaka, Y. Katsumi, N. Takechi, K. Fukae, T. Kato, 2009 J. Electrochem. Soc, 156 H971

29. Y. Miura, Y. Kasahara, H. Habuka, N. Takechi, K. Fukae, 2009 Jpn. J. Appl. Phys.48 026504

30. Y. Chinone, S. Esaki, F. Fujita, 1996 Shirikon no KagakuRealize, Tokyo 890

31. H. Habuka, T. Sukenobu, H. Koda, T. Takeuchi, M. Aihara, 2004 J. Electrochem. Soc.151 G783 787

32. D. R. Lide, H. P. R. Frederikes, 1997 CRC Handbook of Chemistry and Physics, ed., D. R. Lide (CRC Press, Boca Raton, FL, 78 th ed.1

33. T. E. Lee, 1998 A Beginner's Guide to Mass Spectral Interpretation (Wiley, Chichester, 1998).

34. J. R. Chapman, 1993 Practical Organic Mass Spectrometry Wiley, Chichester

35. H. Habuka, T. Otsuka, W. F. Qu, Jpn, 1999 J. Appl. Phys.38 6466

36. Y. Saito, M. Hirabaru, A. Yoshida, 1992 J. Vac. Sci. Technol B, 10 175

37. S. V. Patankar, 1980 Numerical Heat Transfer and Fluid Flow, McGraw-Hill, New York, USA

38. Webbook http://webbook.nist.gov/chemistry

39. R. C. Reid, J. M. Prausnitz, B. E. Poling, 1987 The Properties of Gases and LiquidsMcGraw-Hill, New York 4 th

40. Nihon Gakujutsushinkokai, , 1997 Fussokagaku Nyumon, 86 Nikan Kogyo ShinbunshaTokyo, Japan [in Japanese]

41. Fluent 2001 FLUENT User's ManualVer. 5.5 (Fluent, Inc., Hanover, 2001)

42. H. Matsunami, 2003 Technology of Semiconductor SiC and Its ApplicationNikkan KogyoTokyo, Japan

43. T. Ohno, H. Yamaguchi, S. Kuroda, K. Kojima, T. Suzuki, K. Arai, 2004 J. Cryst. Growth260 209

44. X. Ma, M. Dudley, W. Vetter, T. Sudarshan, 2003 Jpn. J. Appl. Phys., 42 L1077

45. J. Takanashi, M. Kanaya, Y. Fujiwara, 1994 J. Cryst. Growth135 61

4

ELECTROOPTICAL CHARACTERIZATION OF NEW CLASSES OF SILICON CARBIDE UV PHOTODETECTORS

G. Adamo,[1] A. Tomasino,[1] A. Parisi,[1] D. Agrò,[1] S. Stivala,[1] L. Curcio,[1] A. Andò,[1] R. Pernice,[1] C. Giaconia,[1] A. C. Busacca,[1] M. C. Mazzillo,[2] D. Sanfilippo,[2] and G. Fallica[2]

[1]Dipartimento di Energia, Ingegneria Dell'Informazione e Modelli Matematici (DEIM), University of Palermo, 90128 Palermo, Italy

[2]STMicroelectronics, R&D IMS, 95121 Catania, Italy

In this paper, we present the fabrication process steps and the characterization of 4H-SiC vertical Schottky UV detectors, where interdigitated strips, acting as top metal contacts, have been realized in Ni_2Si. These devices exploit the pinch-off surface effect. I-V and C-V characteristics, as functions of temperature, were measured in dark conditions. In addition, we have carried out responsivity measurements, for wavelengths ranging from 200 to 400 nm, at varying package temperature and applied reverse bias. A comparison among devices having different strip pitch sizes has been performed, thus finding out that the 10-μm pitch class demonstrates the top performances as regards the tradeoff between exposed surface area and complete merge of adjacent depleted regions under top contacts.

SECTION I

INTRODUCTION

High signal-to-noise ratio (SNR), high responsivity within a wide spectrum, photocurrent linearly scaling with the incident optical power within the largest possible range, and suitable spectral selectivity are usually some of the most important characteristics of photodetectors. In any event, the relative weight of such features is related to the specific application [1], [2].

Until now, the most frequently used devices for UV light monitoring have been the photomultiplier tubes (PMT) [3], [4]. Unfortunately, they possess some limitations, such as unsatisfactory efficiency in the UV range, bulky size, expensiveness, and high supply voltage. For these reasons, challenge has focused on the fabrication of a more integrated, cost-effective and new technological efforts. Indeed, the increasing demand of portable UV detection systems has led to new investments on semiconductor-based detectors, e.g., photoconductors, PIN diodes, metal semiconductor (Schottky) junction photodetectors.

Recently, the application fields employing UV signals have been increasing in several contexts, such as astronomy, biology and medicine, security, and non-line-of-sight (NLOS) communications[3], [5]–[6] [7] [8] [9] [10] [11] [12][13].

The aforementioned applications depend on the great progress in bulk and epitaxial growth technologies and on the processes available for wide bandgap materials. Among the latter, Silicon Carbide (SiC) has been revealed as suitable semiconductor for high-speed UV detection. Indeed, SiC-based devices present a high quantum efficiency -with a peak value around 290 nm-, low dark current and very good visible blindness.Furthermore, their straightness consists in the capability of working even at a few hundred degrees centigrade[14]–[15] [16] [17][18]. Besides, the recently realized SiC UV photodiodes possess crucial functional characteristics, such as durable thermal resistance under high radiation and the highest performance concerning the sensitivity to low photon fluxes[3], [6], [15], [18]–[19][20].

In this work, we report on the design of three novel types of 4H-SiC UV photodetectors, herein briefly called SiC8, SiC10, and SiC20, that have been fabricated by employing nickel silicide interdigitated strips whose widths are 8, 10, and 20, respectively. We first report on and characteristics and then the comparison among the optical behavior of the three class devices, highlighting the dependence on temperature. We also examine responsivity measurements as a function of the wavelength, of the applied bias and investigate how the manufacturing defects can negatively impact the performances of the devices.

SECTION II

DEVICE FABRICATION

All the SiC devices were produced at the STMicroelectronics R&D facilities, located in Catania, Italy. The process flow is briefly described in the following. An slightly n-type 4H-SiC epitaxial layers, whose thickness is fixed at 4, is grown onto an n-type heavily doped substrate of the same material (homo-junction). The doping concentration of the epitaxial layer and the substrate are and , respectively. The ohmic contacts on the opposite side of the sample were realized by sputtering a 200 nm Ni film and by a following fast annealing process around 1000°C in order to form the Ni-Si bonds. For the previous class of photodiodes, presented in[19]–[20] [21][22], the surface area of the chip was being covered by a thick sacrificial oxide layer, later defined via photolithography and a highly selective etching step in order to realize 2- wide strips. Then, a fast thermal treatment at low temperature (700°C) was being performed to give rise to the Schottky barrier [19]. During such a process, the area between adjacent strips was being kept covered (i.e., shielded) by the oxide layer. Anyway, the latter steps were no longer carried out for the last device manufacturing, because of the technical complexity related to such a process. Indeed, in the new class of devices presented in this paper, the oxide layer was not present. As a consequence, the Ni film was sputtered over the whole active area and then selectively removed via optical lithography in order to define the geometrical path of the strips.

A 1 AlSiCu alloy layer was sputtered on the top left side of the device as pad for the anode contact, obtained via lithography. Finally, a Ti-Ni-Au layer (100 nm/500 nm/50 nm) was sputtered on the back side in order to form the cathode ohmic contact[21], [22].

In Fig. 1(a), the resulting cross section of the photodiode is shown, whereas a microscope image of the chips is illustrated in Fig. 1(b).

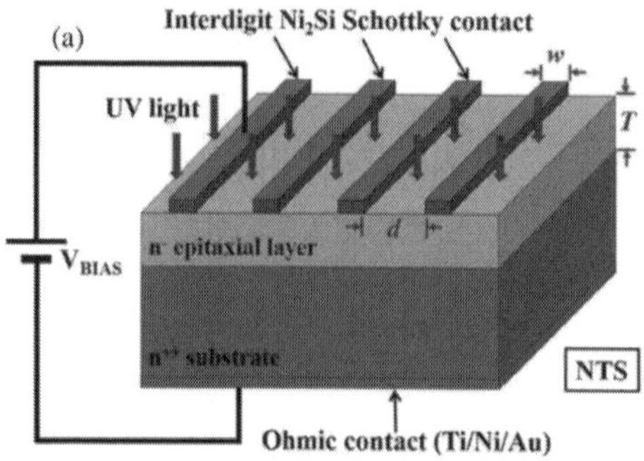

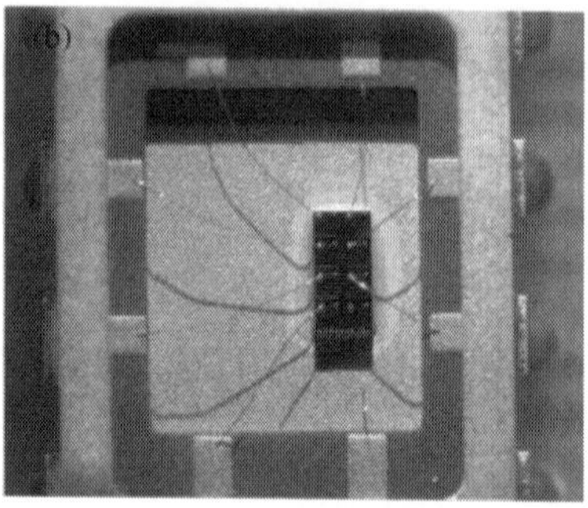

Figure 1. (a) Cross section of the 4H-SiC device. represents the epilayer thickness, is the strip width, and the space among adjacent strips. (b) Optical microscope image of the package including seven photodiodes.

Each photodiode is featured by a top surface and a 2 strip width. The gap between consecutive strips depends on the specific class: 8, 10, and 20, respectively. Thus, the fill factor, i.e., the rate of the top area directly exposed to light is immediately fixed at about 67%, 71%, 81%, respectively.

SECTION III

RESULTS AND DISCUSSION

Herein, we analyze the experimental results of our characterization presented in the previous section. We want to stress the fact that we performed and reported on each measurement at varying temperature thus obtaining a more complete knowledge of the operating conditions of such devices. For the same reason, an in-depth post-processing data has been carried out in order to improve our understanding of the physical behavior of the photodiodes.

3.1. Electrical Characterization

In order to perform our measurements within the temperature range from 20 to 110 °C, we employed a temperature-controlled probe station. The characteristics are plotted in Fig. 2 for SiC8. The leakage current resulted as low as 3 pA at 5 V and 25 °C (room temperature), but it did not change by increasing temperatures. As expected, the threshold voltage decreases by increasing temperature (see Fig. 2(b)). This is due to the presence of more thermally generated carriers. The reverse current (i.e., the dark current) seems to be almost constant and temperature-independent. This means that the thermal energy contributions due to the intermediate levels within the bandgap (which act as recombination centers) could be neglected while they are operating in dark conditions. In order to study the (barrier) capacitance behavior, we employed a precision LCR meter (Agilent 4284A), which applied a 100 kHz readout frequency test signal and a properly tuned reverse bias. It is worth highlighting how the curve in Fig. 2(c) clearly shows a trend switch at a voltage value of about 1 V. Such circumstance represents the first experimental proof that all the depleted regions merge together already at relatively low reverse bias. In more details, such an effect is due to the low doping level of the epitaxial layer (quite close to the intrinsic carrier concentration in SiC). The latter feature allows to create, for reverse biases as low as a few hundred of mV, wide full depleted regions under each strip. This encouraging fact is a strong advantage of our devices, since it makes them suitable in challenging applications where very low or even no power supply is available.

3.2. Optical Characterization

We carried out responsivity measurements within the 200 400 nm spectral range by varying the voltage supply. The set-up illustrated in Fig. 3 was adopted, which is described in more details in [23]–[24][25]. Briefly, a monochromator is employed to filter the incoherent light emitted by a Xenon lamp. The output beam is directed towards the examined sample which is enclosed into a black metallic box. In order to reduce noise, we also employed a lock-in amplifier. Moreover, to achieve a spectral resolution of 2 nm, the selected slits width should be fixed at 1 mm. To direct all the optical power onto the effective area of the device, a lens having a focal length of 20 mm was used. A Peltier cell was employed to control the SiC package temperature which was also recorded by a PT100 thermistor. In order to demonstrate that the devices possess a good spatial uniformity, we analyzed the responsivity at several positions of the sensitive surface keeping constant both the voltage (0 V) and the temperature (25°C).

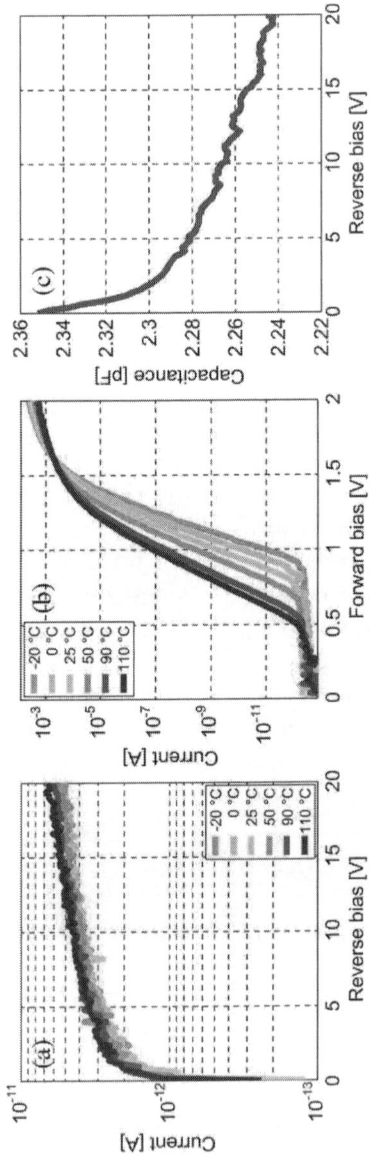

Figure 2. I–V characteristics. (a) Reverse and (b) forward of the SiC8 class, measured within the range and in dark conditions. (c) Capacitance of the SiC8 class at 100 kHz and at a temperature of 25°C, as a function of reverse bias.

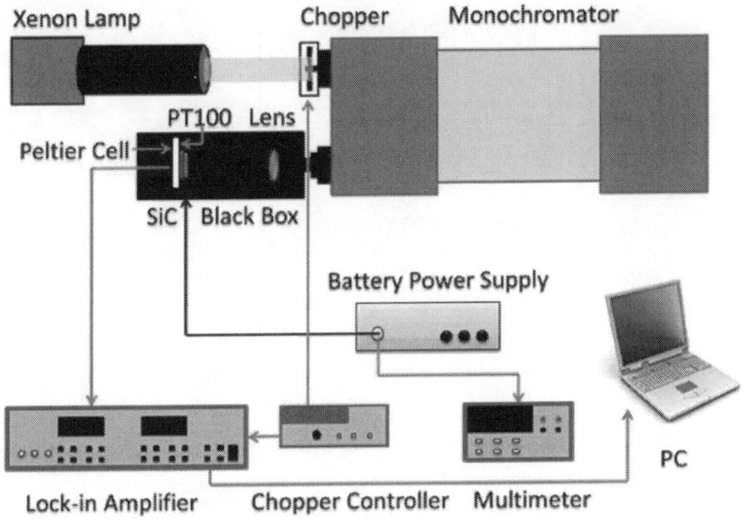

Figure 3. Experimental setup for responsivity measurements.

Fig. 4 illustrates the experimental results performed at three voltages and four temperatures, at the geometrical center of the device area, for the cases of the examined three device classes. It is worth noticing that, at fixed bias voltage, the responsivities weakly vary with temperature within the considered spectral range. Moreover, results show that we obtained an almost full depletion for the SiC8 and SiC10 cases at a , instead, the same was not achievable for SiC20, as the peak responsivity value are lower if compared to the other devices, at any voltage (see also Fig. 5). By using a lock-in amplifier connected to the examined chip, very low power (30 nW) measurements were carried out in all the considered spectrum and also the DC inverse current (mainly due to thermal generation) was rejected.

The light beam was chopped at 183 Hz. A calibrated power meter was employed to detect the incident optical power. The devices showed a considerable photocarriers generation efficiency within the range between 280–300 nm. The change in responsivity is caused by the reverse bias applied (unlike stated in [22]) and by the temperature [26]. We measured the maximum peak value of 0.105 A/W for SiC10 at 290 nm at 0 °C. Furthermore, our responsivity curves point out how the optical response decreases when temperatures increase at 0 V (i.e., the so-called photovoltaic regime in which such types of photodiode commonly operate) while such behavior was less noticeable for greater (3 V and 10 V).

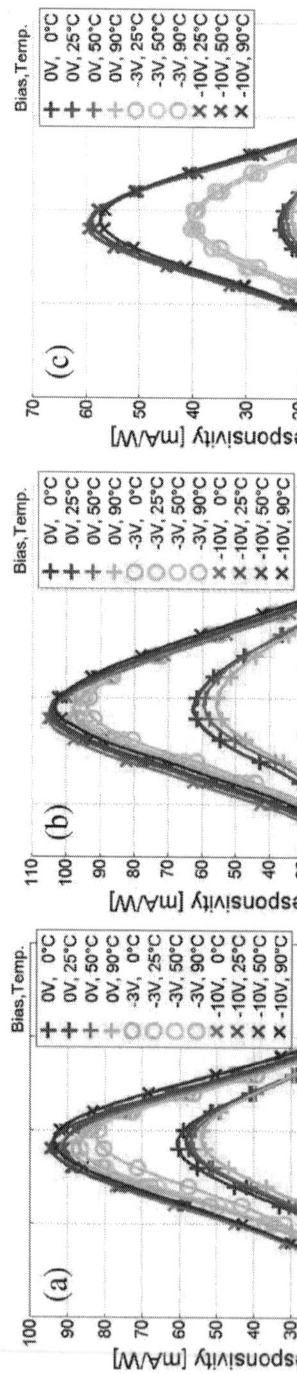

Figure 4. Measurements of SiC8 (a), SiC10 (b), and SiC20 (c) at varying temperatures and reverse biases.

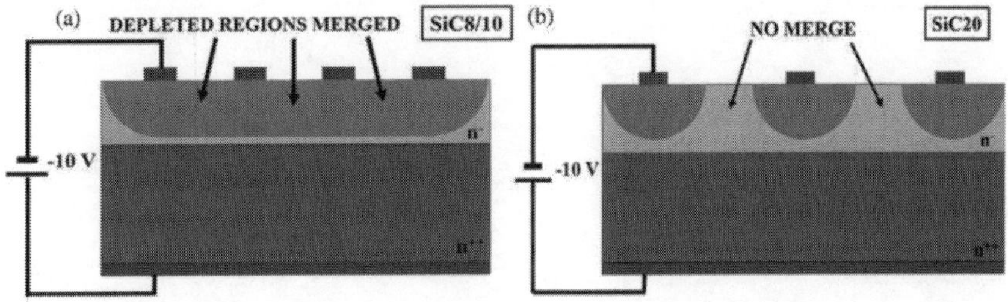

Figure 5. Schematic of the pinch-off effect for SiC8 and SiC10 (a) and for SiC20 (b).

Indeed, the temperature increase could result in a decrease of minority carrier lifetime. In particular, the latter effect could be caused by non radiative recombination centers probably induced during the fabrication process flow, featured by the deep energy levels within the E_g. The presence of such recombination centers could be mainly related to the fabrication process of the strips that is different if compared to the previous class of devices (see Section 2), since the area between adjacent strips is directly exposed to the chamber atmosphere rather than covered by the oxide layer. The carriers could be partially gathered by such centers, and for this reason, they would no longer be involved in photoconduction. These considerations could explain the different behavior with respect to what is reported in [22]. However, deeper investigations are currently ongoing to support this explanation. By increasing (3 V and 10 V), the recombination centers release the captured carriers making the former available for the generation of the photocurrent via indirect transition. This could explain the weak dependence of the responsivity curves on the temperature at the highest reverse bias voltages. Moreover, the decrease of the carrier mobility with temperature affects the responsivity reduction mainly for wavelengths smaller than the peak value (see Fig. 4), since the penetration depth is reduced for the latter, thus resulting in a lower photogenerated carrier collection efficiency.

As expected from theory, the responsivity peak shifts toward slightly longer wavelengths at higher temperatures (see Fig. 6), since the wavelength corresponding to the maximum photon absorption inversely scales to the bandgap of the material (approximately as) that, in turn, shrinks with temperature (, as described by the Varshni's empirical rule[27]).

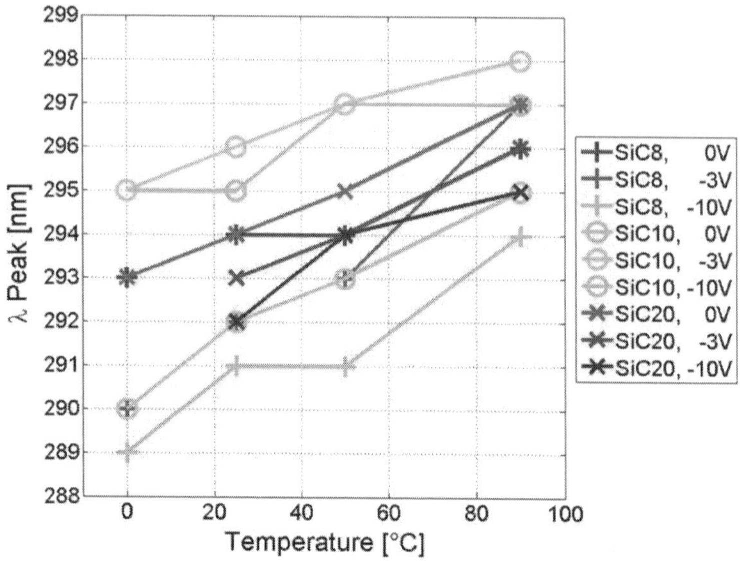

Figure 6. Wavelengths corresponding to the responsivity peaks, obtained at varying reverse biases, on the analyzed device classes as a function of their packages temperature.

SECTION IV

CONCLUSION

In this paper, we have analyzed and compared the properties of three kinds of 4H-SiC vertical Schottky UV photodiodes having (8, 10, and 20 pitch size). Forward and reverse and characteristics at varying temperatures have been traced in dark conditions.

All the three SiC devices showed the best responsivity at 290 nm. Our responsivity measurements highlight that the optical response augments with the applied reverse voltage but decreases with the temperature, especially at 0 V. Furthermore, by increasing the temperature, the optical peak response shifts toward slightly larger wavelengths. Our results show that SiC10 devices presented the best results even at very low bias, which entails the best trade-off in terms of space strip width ratio. Conversely, the SiC20 device showed worse performance not reaching the merge of all the depleted regions for any reverse voltage applied. The obtained results were also analyzed from a physical point of view starting from the afore-mentioned results. Ongoing studies will provide further understanding of these photodiodes thus improving their performances, even in view of the future fabrication of devices exploiting the avalanche effect (APD).

FOOTNOTES

Corresponding author: G. Adamo (e-mail: gabriele.adamo@unipa.it).

REFERENCES

1. G.Giustolisi, G.Palumbo, P.Finocchiaro and A.Pappalardo "A simple extraction procedure for determining the electrical parameters in silicon photomultipliers" *Proc. Eur. Conf. Circuit Theory Design, pp. 1-4, Sep., 2013*

2. G.Giustolisi, R.Mita and G.Palumbo "Behavioral modeling of statistical phenomena of single-photon avalanche diodes" *Int. J. Circuit Theoy. Appl., vol. 40, no. 7, pp. 661-679, Jul., 2012*

3. M.Razeghi and A.Rogalski "Semiconductor ultraviolet etectors" *J. Appl. Phys., vol. 79, no. 10, pp. 7433-7473, May, 2007*

4. M.Mazzillo, A.Sciuto, F.Roccaforte and V.Raineri "4H-SiC Schottky photodiodes for ultraviolet light detection" *Proc. NSS/MIC, pp. 1642-1646, 2011*

5. E.Monroy, F.Omnes and F.Calle "Wide-bandgap semiconductor ultraviolet photodetectors" *Semicond. Sci. Technol., vol. 18, no. 4, pp. R33-R51, Mar., 2003*

6. G. A.Shaw "Deep-UV photon counting detectors and applications" *Proc. SPIE, vol. 7320, pp. 73200J-1-73200J-15, 2009*

7. P.Schreiber "Solar blind UV region and UV detector development objectives" *Proc. SPIE, vol. 3629, pp. 230-248, 1999*

8. D.Engelhaupt, P.Reardon, L.Blackwell, L.Warden and B.Ramsey "Autonomous long-range open area fire detection and reporting" *Proc. SPIE, vol. 5782, pp. 164-175, 2005*

9. A.Tomasino "Wideband THz time domain spectroscopy based on optical rectification and electro-optic sampling" *Sci. Rep., vol. 3, p. 3116, Oct., 2013*

10. A.Pasquazi, A.Busacca, S.Stivala, R.Morandotti and G.Assanto "Nonlinear disorder mapping through three-wave mixing" *IEEE Photon. J., vol. 2, no. 1, pp. 18-28, Feb., 2010*

11. A.Parisi, A. C.Cino, A. C.Busacca, M.Cherchi and S.Riva-Sanseverino "Integrated optic surface plasmon resonance measurements in a borosilicate glass substrate" *Sensors, vol. 8, no. 11, pp. 7113-7124, Nov., 2008*

12. M.Cherchi "Exploiting the optical quadratic nonlinearity of zinc-blende semiconductors for guided-wave terahertz generation: A material comparison" *IEEE J. Quantum Electron., vol. 46, no. 3, pp. 368-376, Mar., 2010*

13. R.Pernice "Opals infiltrated with a stimuli-responsive hydrogel for ethanol vapor sensing" *Opt. Mater. Exp., vol. 3, no. 11, pp. 1820-1833, Nov., 2013*

14. A.Vert, S.Soloviev, J.Fronheiser and P.Sandvik "Solar-blind 4H-SiC single-photon avalanche diode operating in Geiger mode" *IEEE Photon. Technol. Lett., vol. 20, no. 18, pp. 1587-1589, Sep., 2008*

15. S.Metzger, H.Henschel, O.Köhn and W.Lennartz "Silicon carbide radiation detector for harsh environments" *IEEE Trans. Nucl. Sci., vol. 49, no. 3, pp. 1351-1355, Jun., 2002*

16. M.Holz, G.Hultsch, T.Scherg and R.Rupp "Reliability considerations for recent Infineon SiC diode releases" *Microelectron. Reliab., vol. 47, no. 9–11, pp. 1741-1745, Sep.–Nov., 2007*

17. M. A.Borysiewicz "Fundamentals and practice of metal contacts to wide band gap semiconductor devices" *Cryst. Res. Technol., vol. 47, no. 3, pp. 261-272, Mar., 2012*

18. B. K.Ng "Multiplication and excess noise characteristics of thin 4H-SiC UV avalanche photodiodes" *IEEE Photon. Technol. Lett., vol. 14, no. 9, pp. 1342-1344, Sep., 2002*

19. A.Sciuto, F.Roccaforte, S.Di Franco and V.Raineri "High responsivity 4H-SiC Schottky UV photodiodes based on the pinch-off surface effect" *Appl. Phys. Lett., vol. 89, no. 8, pp. 081111-1-081111-3, Aug., 2006*

20. A.Sciuto "On the aging effects of 4H-SiC Schottky photodiodes under high intensity mercury lamp irradiation" *IEEE Photon. Technol. Lett., vol. 22, no. 11, pp. 775-777, Jun., 2010*

21. M.Mazzillo, A.Sciuto, G.Catania, F.Roccaforte and V.Raineri "Temperature and light induced effects on the capacitance of 4H-SiC Schottky photodiodes" *IEEE Sensors J., vol. 12, no. 5, pp. 1127-1130, May, 2012*

22. M.Mazzillo "Highly efficient low reverse biased 4H-SiC Schottky photodiodes for UV-light detection" *IEEE Photon. Technol. Lett., vol. 21, no. 23, pp. 1782-1784, Dec., 2009*

23. G.Adamo "Measurements of silicon photomultipliers responsivity in continuous wave regime" *IEEE Trans. Electron Devices, vol. 60, no. 11, pp. 3718-3725, Nov., 2013*

24. G.Adamo "Responsivity measurements of N-on-P and P-on-N silicon photomultipliers in the continuous wave regime" *Proc. SPIE, vol. 8629, p. 86291A, 2013*

25. G.Adamo "SNR measurements of silicon photomultipliers in the continuous wave regime" *Proc. SPIE, vol. 8990, p. 899 019, 2014*

26. G.Adamo "Responsivity measurements of 4H-SiC Schottky photodiodes for UV light monitoring" *Proc. SPIE, vol. 8990, p. 899 017, 2014*

27. R.Pässler "Semi-empirical descriptions of temperature dependences of band gaps in semiconductors" *Phys. Stat. Sol. (B), vol. 236, no. 3, pp. 710-728, Apr., 2003*

5

STUDY OF TRIBOLOGICAL BEHAVIOR OF SILICON CARBIDE BASED ALUMINUM METAL MATRIX COMPOSITES UNDER DRY AND LUBRICATED ENVIRONMENT

Kalyan Kumar Singh,[1] Saurabh Singh,[1] and Anil Kumar Shrivastava[2]

[1]Department of Mechanical Engineering, Indian School of Mines, Dhanbad 826004, India

[2]Department of Mechanical Engineering, Galgotias College of Engineering & Technology, Greater Noida 201306, India

ABSTRACT

Friction and wear behavior of silicon carbide based aluminum metal matrix composite and aluminum matrix alloy have been studied for sliding speeds of 3.14 m/s and 3.77 m/s and load range from 10 N to 30 N under dry and lubricated environment, respectively. The experiments were performed on pin on disk tribometer (Make: DUCOM). The composite was fabricated by stir casting process and has several challenges like inferior bonds and interfacial reaction products which will deteriorate the mechanical and tribological properties. Therefore, addition of reactive metal like magnesium (Mg) should be done which will lead to reduced solidification shrinkage, lower tendency towards hot tearing, and faster process cycles. Results have revealed that the developed composites have lower coefficient of friction and wear rates when compared with aluminum matrix alloy under dry and lubricated environment. Experimental results show that under dry condition coefficient of friction of both the matrix alloy and the composite decreases with increase in load, whereas it increases with increase in sliding speeds; on the other hand wear rates of both aluminum matrix alloy and the composites increase with increase in load as well as with sliding speeds. FESEM of worn surfaces are also used to understand the wear mechanisms.

1. INTRODUCTION

Aluminum alloy possesses excellent properties like high strength-to-weight ratio, corrosion resistance, good ductility, low density, high thermal conductivity, and so forth, thus making it suitable for wide range of applications in several industries like automobile, aerospace, structural applications, and so forth [1]. In recent times, the uses of high strength aluminum alloys are focused on structural applications in aerospace as well as on general engineering sectors. However aluminum alloys have not been well known for high performance tribological application because of their low hardness, poor wear resistance, and poor weldability. Thus, there is a greater possibility of oscillation since these materials are joined by riveting, bolting, and so forth and due to this it may often lead to sliding wear in dry environmental conditions. Therefore, to overcome this problem reinforcement of particulates, fiber, or whiskers can be used; these reinforcements are very well known for their high-specific strength and uniform distribution [2] which leads to use of metal matrix composites to complete the demand for light weight, high performance, environment friendly, wear and corrosion resistance material. Generally, Ag, Al, Be, Co, Cu, Fe, Mg, Ni, and Ti are produced and used as matrices in metal matrix composites. Due to excellent physical, mechanical, and tribological properties and good dimensional stability of aluminum matrix composite it has high commercial application compared to other available metals. The demand for aluminum composite in automobile sector has been increased drastically; this composite is used for the manufacturing of pistons, brake drums, cylinder liners, and pushrods [3, 4]. A commercial Al 7075 aluminum alloy is one of the highest strength aluminum alloys available and is having excellent strength-to-weight ratio and is formed in annealed conditions and subsequently heat treated. But due to its low wear resistance and poor tribological behavior it cannot be widely used; therefore reinforcement of particles has been done to improve its tribological properties. There are various particles available which are used for reinforcement such as B_4C, SiC, TiC, B, C, and Al_2O_3. With the reinforcement of high modulus ceramics particulates such as silicon carbide (SiC) having properties like more stability and chemical compatibility with aluminum matrix and low cost, wide range of available grade made aluminum alloys are suitable candidate for wider spectrum of applications [5]. The hard reinforcement of silicon carbide particle improved the properties like hardness and wear resistance of composite by acting as load bearing components [6, 7]. Many investigators have studied both parameters, that is, reinforcement size and volume fraction; Aiguo and Rack [8] observed that abrasive wear resistance in unreinforced 7091 alloys is less in comparison with both silicon carbide particle and silicon carbide whiskers. Wear resistance increases and plastic deformation decreases with the increase of reinforcement [9, 10]. Kwok and Lim [11] observed the size of silicon carbide reinforcement at high speed and suggested that when the particulate size becomes small the wear resistance rapidly deteriorates. Therefore, silicon carbide based aluminum metal matrix composites are mainly preferred for low speed applications; silicon carbide is commonly used among the reinforcement particles like carbides, nitrides, borides, and oxides, because it has several advantages like high melting point, high elastic modules, hardness, good thermal conductivity, high strength, excellent wear resistance, good availability, and low cost. Silicon carbide based aluminum metal matrix composites are widely used in automotive, marine, and other sectors because of its excellent properties such as good fluidity and low coefficient of thermal expansion and its quality depends on chemical composition, melting process, casting process, and solidification rate. As the volume fraction addition of silicon carbide in aluminum alloys increases, the value of ultimate tensile strength (UTS) increases and low degree of porosity is obtained.

There is not much improvement in strength and porosity of material, if the volume fraction of silicon carbide exceeds 10%; therefore in the present analysis 10% silicon carbide has been added in aluminum alloys as reinforcement. Due to low degree of porosity and uniform distribution of silicon carbide particles there is enhancement in tensile strength of composites and as a result it leads to effective transfer of tensile loads to the uniformly distributed silicon carbide particulates. In silicon carbide based aluminum metal matrix composites, casting properties increased by adding silicon whereas magnesium (Mg) made its heat treatable which greatly influenced the mechanical properties of composites.

There are several manufacturing techniques that are available for particle reinforcement in metal matrix composites such as liquid infiltration, spray decomposition, squeeze casting, compocasting, powder metallurgy, and mechanical alloying. Among all these available methods, stir casting method is attractive method for manufacturing of metal matrix composites (MMCs) because of low cost, near-net shape formation, and wide range of processing applications [1, 5, 12]. The mechanical strength of aluminum metal matrix composites is greatly influenced by particles size reinforcement; if size of particles ranges from micrometer to nanometric scale there is drastic increase in mechanical strength of composites but it will increase chance of agglomeration and particle clustering which will reduce the strength of composites. Therefore, it is important that reinforcement of particles should be homogeneous to achieve the improved properties of composites [13]. In stir casting method, ceramics reinforcement (i.e., silicon carbide) and metal matrix which is melted during casting are mixed homogeneously by mechanical stirrer. Fabrication of aluminum metal matrix composites has several challenges like inferior bonds and interfacial reaction products and it will directly deteriorate the mechanical properties and tribological properties of composites, which make them unsuitable for industrial application. Therefore, to overcome this problem, addition of reactive metal like magnesium has been done and the addition of this metal results in reduced solidification shrinkage, lower tendency towards hot tearing, suppression of segregation, settling or agglomeration, and faster process cycles [12].

Thus by analyzing various parameters and their effects and utilization this experiment was carried out under various environmental conditions to produce a material which can sustain a longer time. Composite fabricated has much better properties as compared with its base alloys (details and validation of which are shown in the experimental analysis). The severity of worn-out surfaces (i.e., wear scars, change of microstructure) was also analyzed using field emission scanning electron microscopes (FESEM).

2. EXPERIMENTAL PROCEDURE

2.1. Fabrication of Silicon Carbide Based Aluminum Metal Matrix Composite

The composite was prepared by commercially available aluminum (Al 7075) matrix alloy and silicon carbide (10% by weight) which is used as reinforcement. The average size of aluminum matrix alloy was 150 microns to 160 microns and that of silicon carbide was 40 microns to 50 microns. In the present investigation, aluminum matrix alloy was chosen as a base matrix since it is having very high tensile strength of about 275 MPa. The addition of silicon carbide will enhance the wear resisting property of silicon carbide based aluminum metal matrix composite. The chemical composition of aluminum matrix alloys is presented in Table 1.

2.2. Composite Preparation

In order to achieve high level of mechanical properties of the composite, a good interfacial bonding (wetting) between the dispatched phase and the liquid matrix has to be obtained. Fabrication of composite with discontinuous fiber can be economically done using liquid metallurgy route/technique [10]. Stir casting is a liquid state method which is the simplest and is cost effective to fabricate metal matrix composite. Care was taken to maintain optimum casting parameters of pouring temperature (670°C) stirring speed of 500 rpm, stirring time 20 minutes. The reinforcement was preheated at 600°C to the addition in aluminum matrix alloy (Al 7075). To avoid moisture from the particles, preheating of the reinforced particle should be done; otherwise, there is a chance of agglomeration of particulate which occurs due to moisture and gasses present in the particles. Degassing agent (hexachloroethane and magnesium) was used to reduce the gas porosities. The addition of magnesium will also lead to reduced solidification shrinkage, lower tendency towards hot tearing, suppression of segregation, settling or agglomeration, and faster process cycles. The depth of immersion of stirrer was maintained at about two-thirds of the depth of molten metal. The molten metal was then poured into a permanent mould preheated at 300°C of 9 mm diameter and 55 mm length. Specimen was left for curing and die was released after 6 hours; finally, on the lathe cast specimens were parted out in 2 pieces, that is, 25 mm length and 8 mm diameter, respectively. The chemical compositions of silicon carbide based aluminum metal matrix composite are presented in Table 2.

Table 1. Chemical composition (weight %) of aluminum matrix alloy (Al 7075).

Material	Si	Fe	Cu	Cr	Zn	Ti	Mn	Mg	Al	Others
Content	Max 0.4	0.5	1.2–2	0.18–0.28	5.1–6.1	0.2	0.3	2.1–2.9	87.1–91.4	0.15

Table 2. Chemical composition of silicon carbide based aluminum metal matrix composite.

Material	C	O	Mg	Al	Si	Fe	Ag	Zn	Cu
Content	24.27	20.61	1.13	36.43	10.62	1.96	0.84	3.34	0.80

2.3. Composite and Aluminum Matrix Alloy Testing Procedure

The experiment was conducted at room temperature on a pin on disk wear and friction maintaining machine (Make: DUCOM instruments private limited, Bangalore diameter of 165 mm, thickness of 8 mm and made of material EN31 hardened to 60 HRC ground to 1.6 Ra surface roughness. Figure 1 shows the photographic view of disk). The size of cylindrical specimen which was used for testing is 8 mm diameter as well as 25 mm length. Before performing every experiment disk surface was cleaned by rubbing 600-grade Si6 emery paper and washed with acetone. Three different loads, that is, 10 N, 20 N, and 30 N, have been selected for wear and friction behavior of specimen with track diameter of 100 mm and 120 mm used for both aluminum matrix alloy and silicon carbide based aluminum metal matrix composite. The test duration was 10 minutes at a constant speed of 600 rpm for the entire test in dry and oil lubricated (SAE 20 with kinematic viscosity 25 to 30 centistokes at 50°C and oil has been used for lubrication) conditions. For oil lubricated experiments, initially, 2 drops of lubricating oil were placed over sliding surface after that oil was poured at 0.02 mL/min. All sets of experiments were repeated three times in the same manner for dry and oil lubricated conditions, respectively. Frictional forces and coefficient of friction were measured directly from the machine. For the calculation of wear loss in terms of weight, weights of specimen before and after the experiment were taken.

The sliding speed of the test specimen is calculated by using the following formula:The following calculation is used to obtain wear rate:

3. RESULTS AND DISCUSSION

3.1. Coefficient of Friction

The variations of coefficient of friction for aluminum matrix alloy (Al 7075) and composites under dry and lubricated condition at different normal loads and two sliding speeds (i.e., 3.14 m/s and 3.77 m/s) have been shown in Tables 3 and 4, respectively. Table 3 shows that there is significant decrease in the value of coefficient of friction for composite compared with aluminum matrix alloy under dry and lubricated condition. Under dry condition, 30%–55% reductions in value of coefficient of friction for load range 10 N–30 N have been observed; this is because of uniform mixing of silicon carbide with aluminum matrix alloy and results in excellent bond between matrix alloy and silicon carbide particle. These reinforced silicon carbide particles act as load bearing members as the loads increase and reduce the touch area between specimen and rotating disk's surfaces [1] which leads to lower coefficient of friction. Similar observations have been observed by other investees [14, 15].

Figure 1. DUCOM Machine.

However, under lubricated condition, the value of coefficient of friction reduced for both aluminum matrix alloy and composites compared with dry condition; this is because of thin lubricating film formation between specimen and rotating disk surface. This thin lubricating film contains lumps of particles and as temperature and normal loads increase these particles break into a small number of particles and reorient themselves along the sliding plane and this leads to lower coefficient of friction.

Table 3. Variation of coefficient of friction with load for aluminum matrix alloy and silicon carbide based aluminum metal matrix composite at = 3.14 m/sec.

Sliding condition & load	COF for aluminum matrix alloy	COF for silicon carbide based aluminum metal matrix composite
Dry 10 N	0.65	0.42
Dry 20 N	0.57	0.37
Dry 30 N	0.47	0.30
Lubricated 10 N	0.46	0.32
Lubricated 20 N	0.43	0.25
Lubricated 30 N	0.36	0.19

Track diameter = 100 mm, $N = 600$ rpm, sliding speeds = 3.14 m/sec, and $\Delta T = 600$ sec.

Table 4 shows variation of coefficient of friction at increased sliding distance for both aluminum matrix alloy and composites. As the sliding distance increases the value of coefficient of friction increases for both aluminum matrix alloy and composites because of temperature rise of rubbing surfaces for aluminum matrix alloy and reinforcement of silicon carbide into aluminum matrix alloys makes the matrix alloy plastically constrained and it can sustain high temperature compared with aluminum matrix alloy. Silicon carbide also protects the matrix alloy from severe contact with counter surface and leads to lower coefficient of friction. Similar observations have been made by other investees [16]. As sliding distance increases, the thin lubricating film between specimen and rotating disk surface may get thickened and be responsible for breakage of small particles from specimen and these broken particles may be clogged between the rubbing surfaces and this leads to higher coefficient of friction.

Table 4. Variation of coefficient of friction with load for aluminum matrix alloy and silicon carbide based aluminum metal matrix composite at = 3.77 m/sec.

Sliding condition & load	COF for aluminum matrix alloy	COF for silicon carbide based aluminum metal matrix composite
Dry 10 N	0.72	0.56
Dry 20 N	0.63	0.43
Dry 30 N	0.50	0.35
Lubricated 10 N	0.58	0.44
Lubricated 20 N	0.47	0.32
Lubricated 30 N	0.37	0.25

Track diameter = 120 mm, N = 600 rpm, sliding speeds = 3.77 m/sec, and ΔT = 600 sec.

Figures 2 and 3 show the variation of coefficient of friction with increasing normal loads under dry and lubricating condition at sliding speeds 3.14 m/s and 3.77 m/s, respectively. It has been observed that temperature rise is greater in case of unreinforced aluminum matrix alloys compared to those in composites irrespective of the applied load and surface conditions. The rise in temperature leads to frictional heating in between the rubbing surfaces and this frictional heating is continuous because there is much less time for heat dissipation that causes higher coefficient of friction for aluminum matrix alloy compared to composites. Initially, aluminum alloys and composites particles are stronger and sharper which causes higher frictional heating and leads to higher coefficient of friction but as the time increases, there is increase in flow ability of material on specimen surface that leads to increase in slipping action and results in reduced frictional heating that leads to lower coefficient of friction at higher loads [1]. The composite shows lower coefficient of friction compared with aluminum matrix alloy because of presence of secondary phase of silicon carbide particles in aluminum matrix alloy which will restrict the flow of metal during sliding [17]. The uniform distribution of silicon carbide particle in aluminum matrix alloy would lead to sustaining higher loads with less plastic deformation and give us good load bearing capacity of composite at higher loads. The minimum values of coefficient of friction for both aluminum matrix alloy and composite have been observed at lower speeds; the same observations have been made by other investees [11].

3.2. Wear Studies

3.2.1. Weight Loss and Wear Rates

Effects of increasing loads (i.e., 10 N, 20 N, and 30 N) with sliding distance 1884.96 m (numerical values shown in Table 5) and 2262 m (numerical values in Table 6) on weight loss and wear rates for both aluminum matrix alloy and composites under dry and lubricated condition have been shown in Figures 4 and 5, respectively.

Table 5: Variation of weight loss and wear rate for aluminum matrix alloy and silicon carbide based aluminum metal matrix composite when sliding speeds = 3.14 m/s.

Table 6: Variation of weight loss and wear rate for aluminum matrix alloy and silicon carbide based aluminum metal matrix composite when sliding speeds = 3.77 m/s.

Figure 4 shows the variation of weight loss and wear rates at sliding distance 1884.96 m with increasing loads. Mazahery and Shabani [18] observed that wear rates vary with normal rate and significantly it is low for composites compared to unreinforced alloys and it is indicative of Archard's law. In dry condition, rate of weight loss increases with increase in loads for both aluminum matrix alloy and composites because, with increasing loads, the rate of abrasion also increases and it will lead to detachment of metals from surfaces and leads to more loss of materials. But the rates of weight loss in composites are lower compared with aluminum matrix alloy because the presence of silicon carbide in aluminum matrix alloy controls the deformation and leads to less weight loss. The worn surfaces of composites are smoother and also shallower groove formations as the time increases which lead to higher wear resistance for composites compared with unreinforced aluminum matrix alloys [5]. As earlier we have seen that wear rates are directly proportional to weight loss, therefore wear rates will also be lower for composites compared with aluminum matrix alloy.

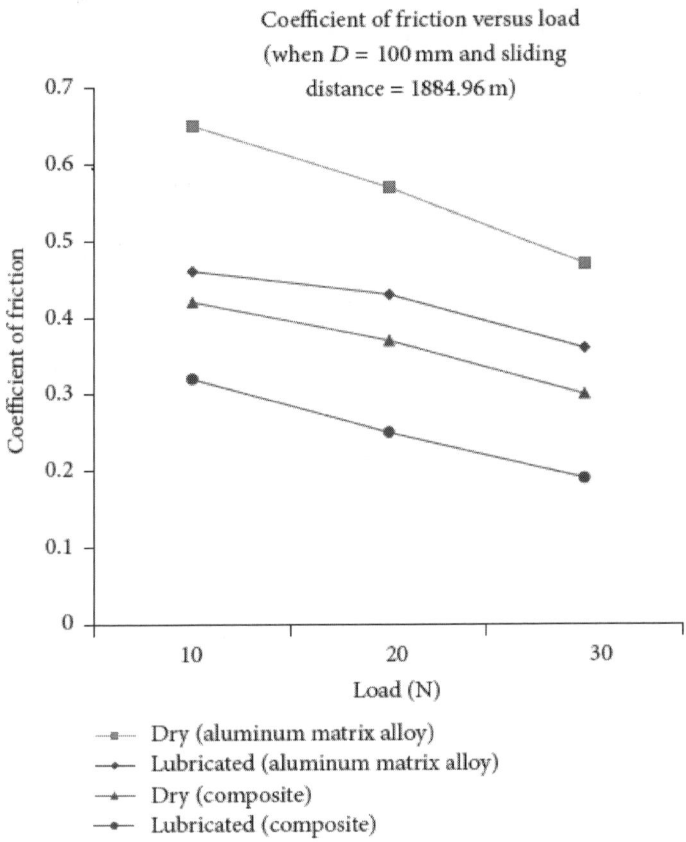

Figure 2. Variation of COF at m/s.

In lubricated condition, wear rates are lower compared with dry condition for both aluminum matrix alloy and composites because of thin lubricating film formation over rotating disk surfaces, and this thin lubricating film decreases the adhesion of small broken particles of aluminum matrix alloy and composites and leads to lower wear rates although it increases with increasing loads and if an applied load exceeds critical values, reinforced particle will break and loses its load carrying capacity properties and leads to more wear rates [5]. Critical loads basically mean that wear rates increase marginally with linear trends but suddenly these trends will not follow the loads up to which these trends that follow are known as critical loads.

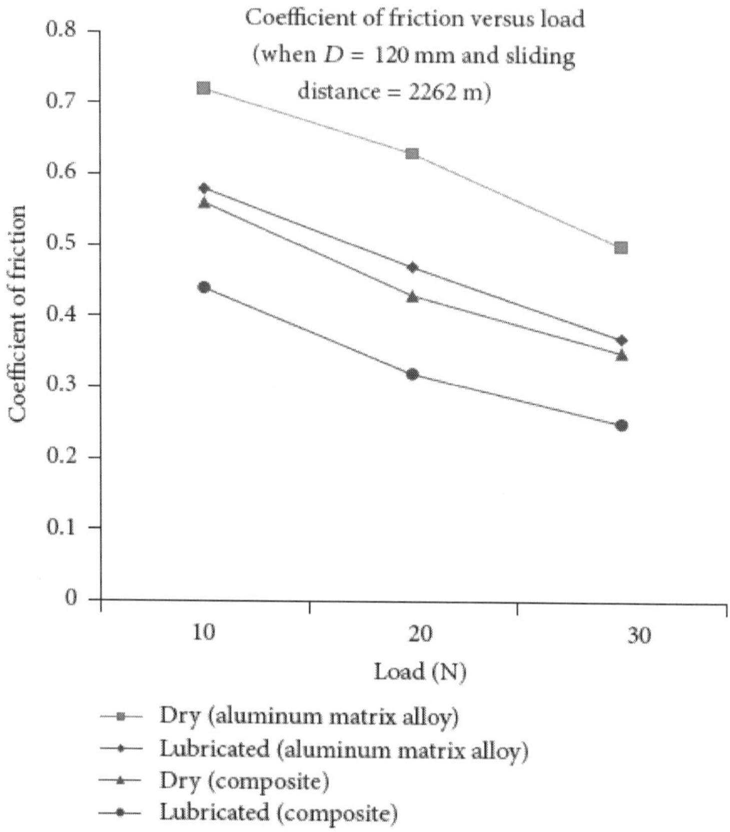

Figure 3. Variation of COF at m/s.

Figure 5 shows the variation of weight loss and wear rates with increased sliding distance 2262 m with increasing loads. The increase of sliding distance leads to increase in weight loss and wear rates for both aluminum matrix alloy and composites because as sliding distance increases, the contact between rotating disk surface and sliding surface of specimen becomes more familiar with elapse of time; because of this the temperature between the rotating disk surface and sliding surface of specimen increases and leads to softening of materials and plastic state of materials occurs. Similar observations have been made by other investees [19,20].

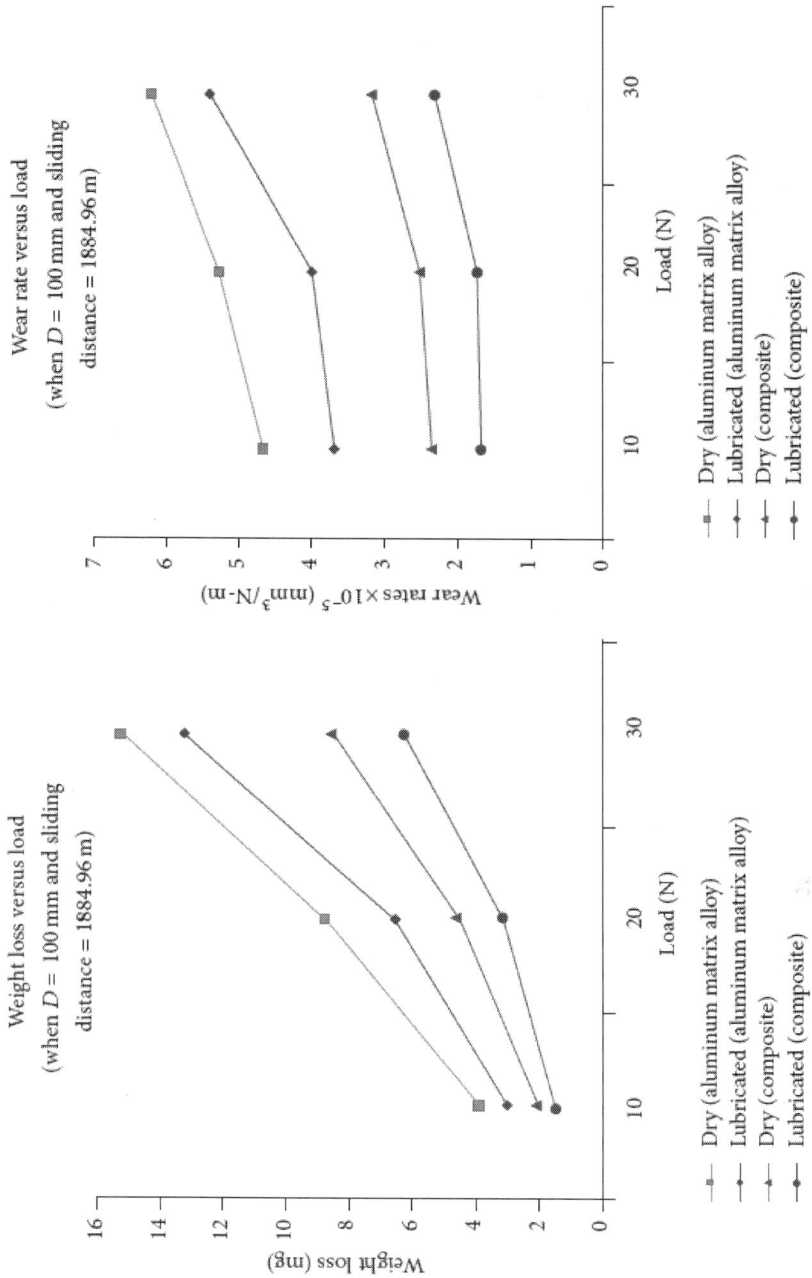

Figure 4. Variation of weight loss and wear rates at sliding distance 1884.96 m with increasing loads.

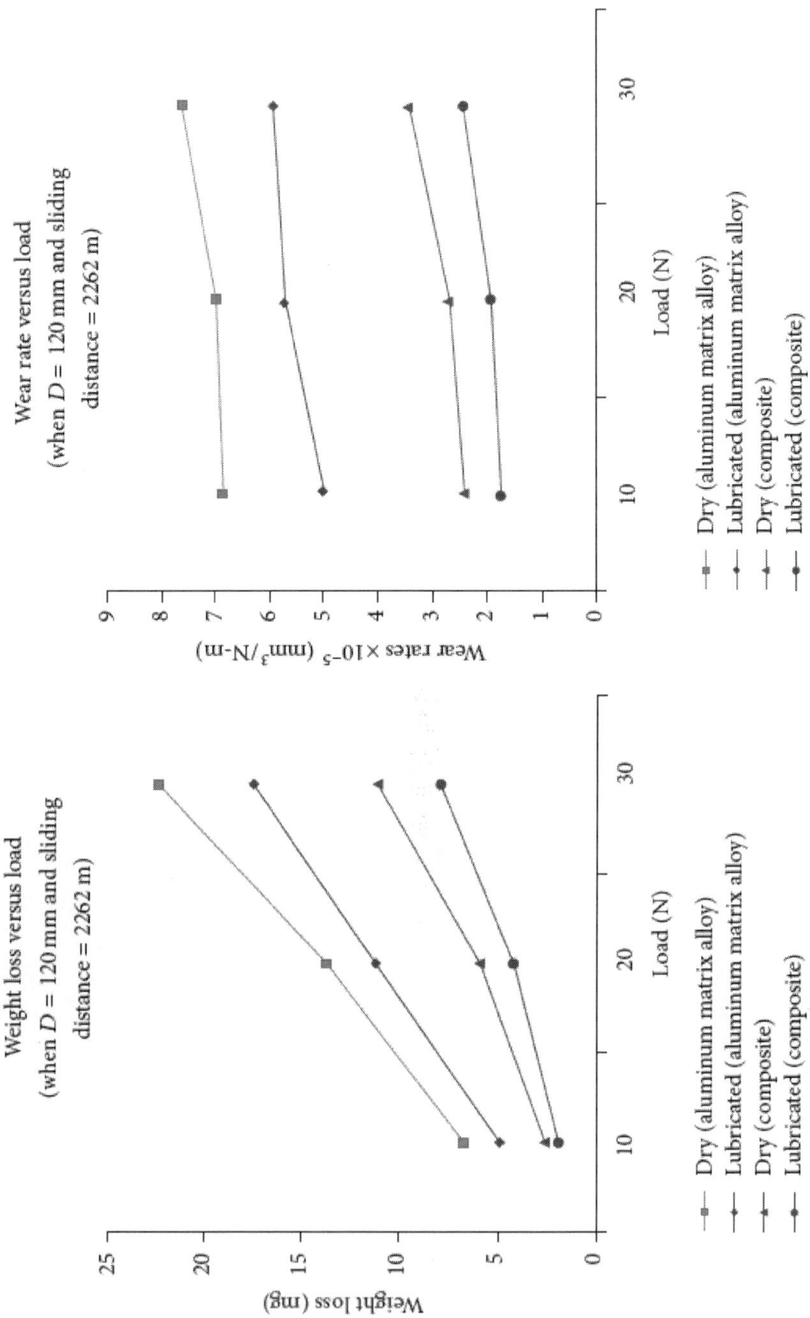

Figure 5. Variation of weight loss and wear rate at sliding distance 2262 m with increasing loads.

3.2.2. FESEM Analysis

FESEM micrographs of all worn surfaces for both aluminum matrix alloy and composites under dry and lubricated condition at sliding speeds 3.14 m/s and 3.77 m/s have been shown at magnification rate (×500).

Figure 6 shows the FESEM analysis of all worn surfaces for both aluminum matrix alloy and composites at different loads with sliding speeds 3.14 m/s under dry condition. It is clearly observed that, because of plastic flow of material, the wear rates for both aluminum matrix alloy and composites increase with increase of loads and severity of plastic flow increases at higher loads. At higher loads, the wear debris is mixture of fine and coarse size particles; this is because of breakdown of large size particles. As the load increases the intersection between the sliding surface of specimen and rotating disk surface increases and leads to breakage of large particles into smaller number of particles. At lower loads, worn surfaces are smooth and grooves are small in size compared with higher loads; with increase in loads these grooves become severe and lead to formation of fracture spots. At these conditions reinforced particles cannot be in stable equilibrium and this leads to increase in wear rates.

Figure 7 shows the FESEM analysis of all worn surfaces for both aluminum matrix alloy and composites at different loads with sliding speeds 3.14 m/s under lubricated condition. In lubricated condition lower wear rates have been observed compared with dry condition because of thin lubricating film formation between sliding surface of specimen and rotating disk surface and it decreases the ploughing action and leads to decrease in breakage of small particles from aluminum matrix alloy and composites; therefore, lower wear rates have been observed in lubricated condition compared with dry condition.

Figure 8 shows the FESEM analysis of all worn surfaces for both aluminum matrix alloy and composites at different loads with sliding speeds 3.77 m/s under dry condition. As the sliding distance increases, there is gradual increase in wear rates for both aluminum matrix alloy and composites. The above results can be observed because of the following reason:(i)With increase in sliding distance, the temperature at specimen interface increases and leads to softening the materials and plastic flow of materials.(ii)With increase in sliding distance, the contact between sliding surface of specimen and rotating disk surface becomes intimate as time increases resulting in higher wear rates.However, the wear rates are lower for composites compared with aluminum matrix alloy because some particles of silicon carbide break into small pieces and crushed as the disk rotates and act as abrasive medium. These powdered silicon carbide particles along with aluminum small particles made a layer which is called mechanically mixed layer (MML). These layers prevent the plastic flow and reduce the wear rates compared with aluminum matrix alloy although plastic flow of composites occurs at later stage.

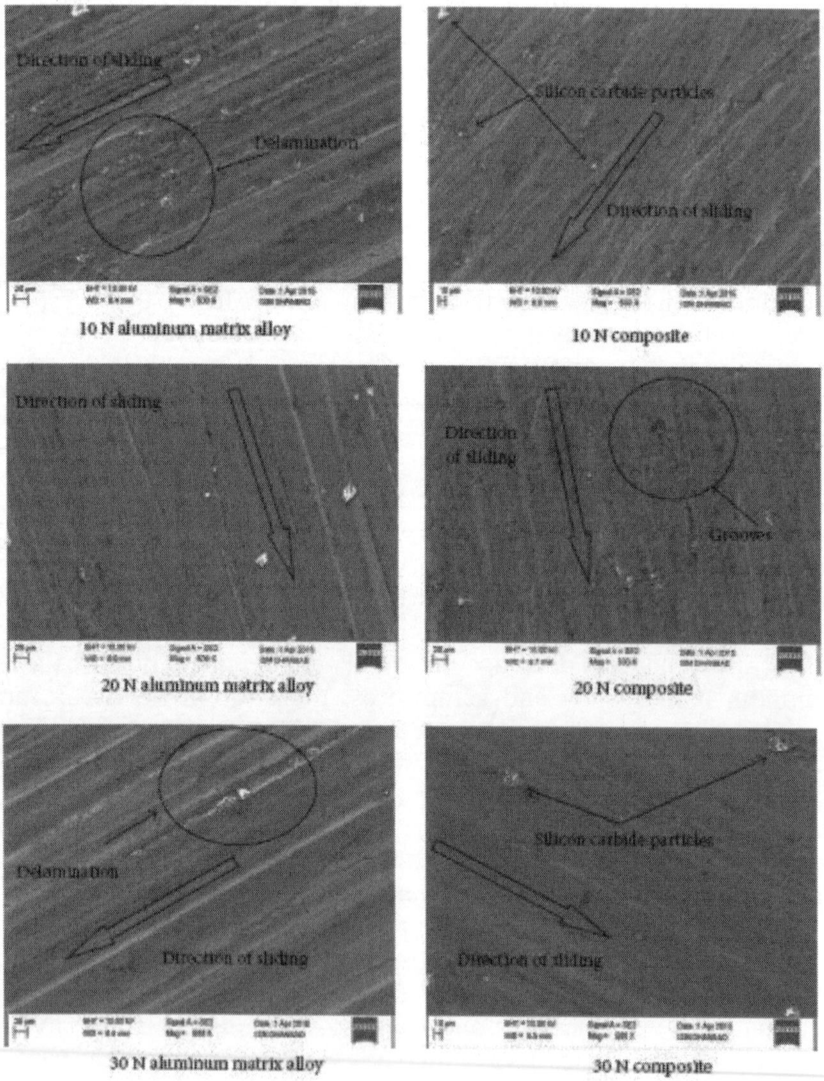

Figure 6. FESEM of worn surfaces at different loads with sliding speeds 3.14 m/s under dry condition.

Figure 9 shows the FESEM analysis of all worn surfaces for both aluminum matrix alloy and composites at different loads with sliding speeds 3.77 m/s under lubricated condition. At higher sliding speeds, wear rates are lower in lubricated condition because of thin lubricating film formation and another mechanically mixed layer (MML) formation delays the plastic flow of material and leads to lower wear rates in lubricated condition compared with dry condition.

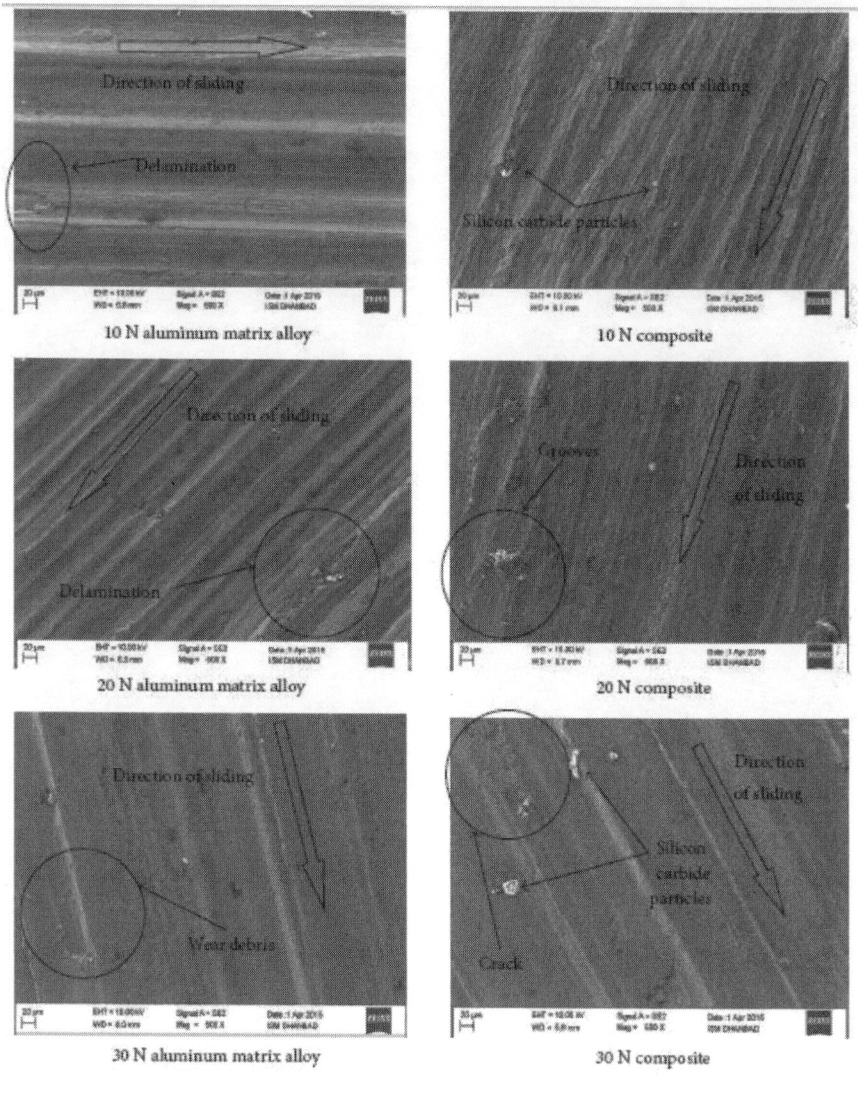

Figure 7. FESEM of worn surfaces at different loads with sliding speeds 3.14 m/s under lubricated condition.

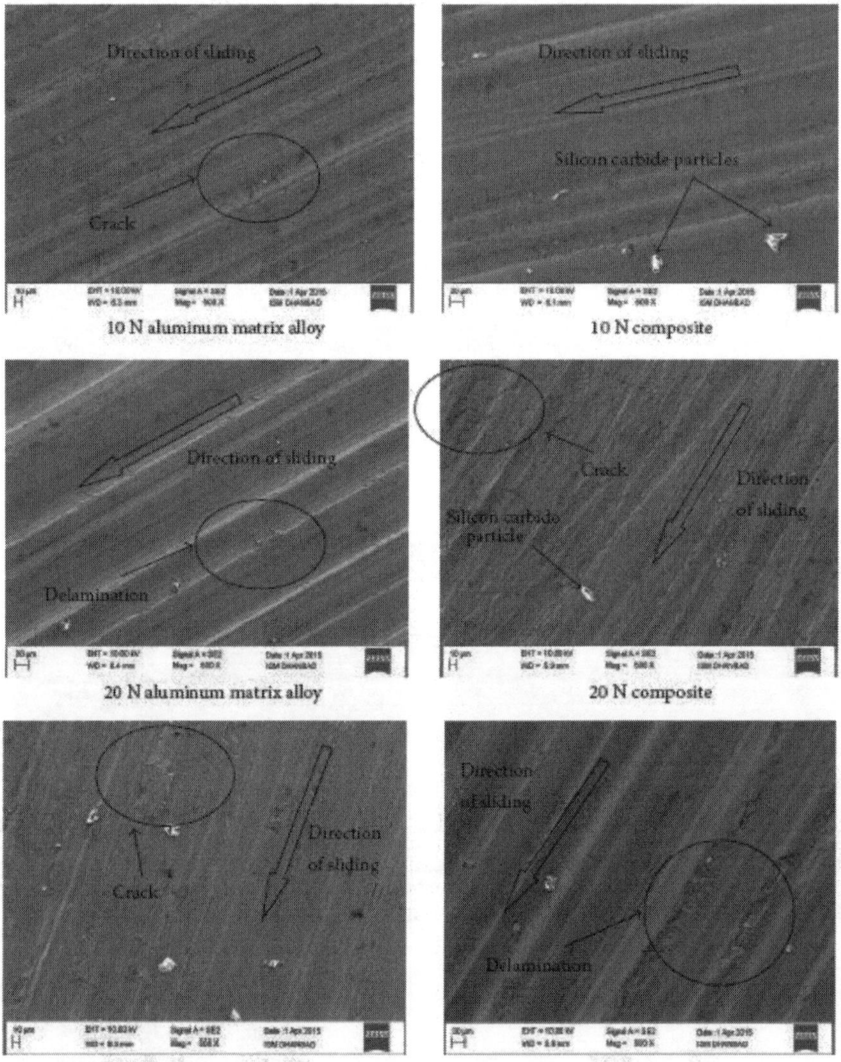

Figure 8. FESEM of worn surfaces at different loads with sliding speeds 3.77 m/s under dry condition.

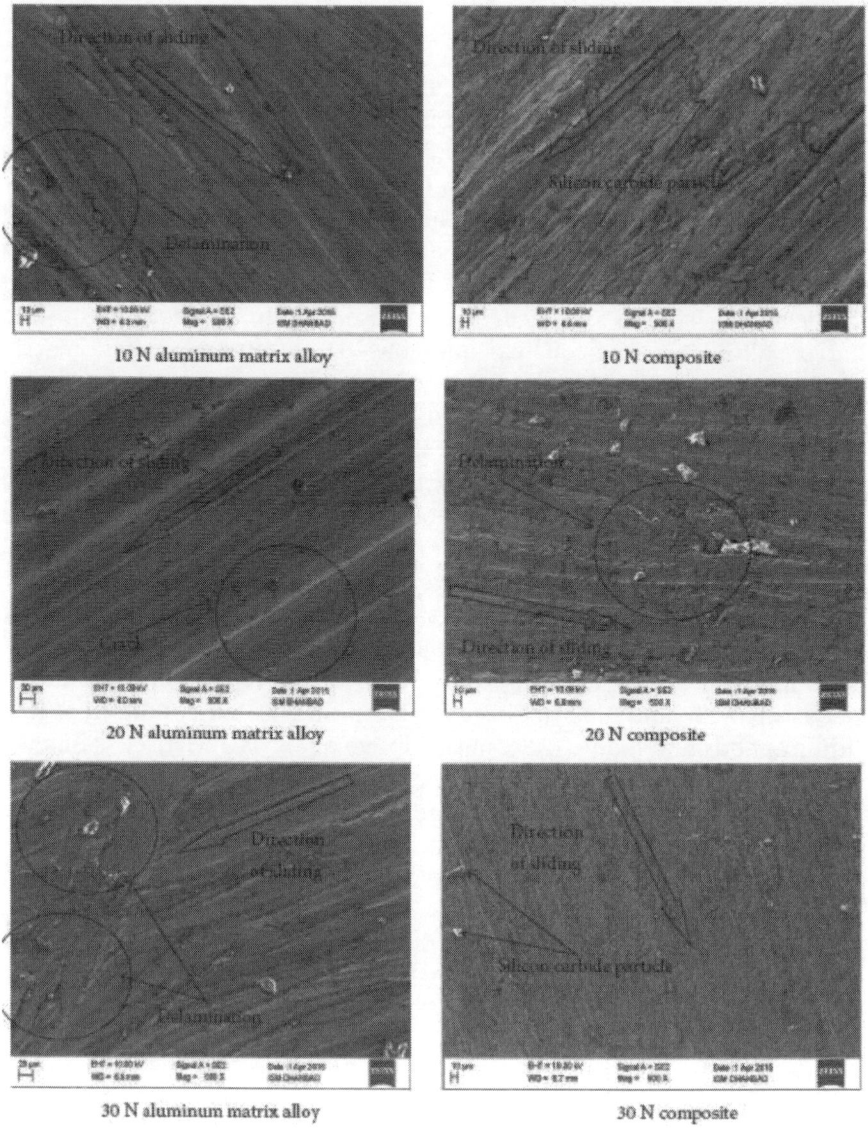

Figure 9. FESEM of worn surfaces at different loads with sliding speeds 3.77 m/s under lubricated condition.

4. CONCLUSIONS

After observing the condition, that is, dry and oil lubricated condition of aluminum matrix alloy and silicon carbide based aluminum metal matrix composite, the following conclusion can be drawn:(1)The values of coefficient of friction are found to be maximum in case of dry condition in matrix alloy at sliding speed 3.77 m/s and minimum in case of lubricated condition in composite at sliding speed 3.14 m/s. The range of coefficient of friction lies from 0.19 to 0.72.(2)The values of coefficient of friction and wear rates are minimum for both aluminum matrix alloy and silicon carbide based aluminum metal matrix composites under lubricated condition compared with dry condition.(3)As we increase normal load and sliding speed, the wear rate increases and it is found to be maximum in dry condition of matrix alloy at 30 N.(4)Uniform distribution of silicon carbide in aluminum matrix alloy reduces the values of coefficient of friction and wear rates for composites compared with aluminum matrix alloy under dry and lubricated condition which prolongs the life of composites for longer duration.(5)The reinforcements of silicon carbide in aluminum matrix alloy significantly improve the strength of material in dry and lubricated condition and act as load bearing members as the loads and sliding speeds increase and give the best results in lubricated condition at low sliding speeds and loads.(6)From the above analysis and discussions we conclude that silicon carbide based aluminum metal matrix composite should be preferred for low speed applications as the same results are also observed by other researchers [21].

ACKNOWLEDGMENTS

The authors are grateful to the Advanced Composite Lab of Department of Mechanical Engineering of Indian School of Mines, Dhanbad, India, for providing them with the experimental support for carrying out the laboratory work.

REFERENCES

1. A. Mazahery and M. O. Shabani, "Study on microstructure and abrasive wear behavior of sintered Al matrix composites," Ceramics International, vol. 38, no. 5, pp. 4263–4269, 2012.

2. C. G. E. Mangin, J. A. Isaacs, and J. P. Clark, "MMCs for automotive engine applications," JOM, vol. 48, no. 2, pp. 49–51, 1996.

3. G. Straffelini, F. Bonollo, A. Molinari, and A. Tiziani, "Influence of matrix hardness on the dry sliding behaviour of 20 vol.% Al_2O_3-particulate-reinforced 6061 Al metal matrix composite," Wear, vol. 211, no. 2, pp. 192–197, 1997.

4. H.-S. Chu, K.-S. Liu, and J.-W. Yeh, "An in situ composite of Al (graphite, Al4C3) produced by reciprocating extrusion," Materials Science and Engineering A, vol. 277, no. 1-2, pp. 25–32, 2000.

5. A. Mazahery and M. O. Shabani, "The effect of primary and secondary processing on the abrasive wear properties of compocast aluminum 6061 alloy matrix composites," Protection of Metals and Physical Chemistry of Surfaces, vol. 50, no. 6, pp. 817–824, 2014.

6. S. C. Lim, M. Gupta, L. Ren, and J. K. M. Kwok, "The tribological properties of Al–Cu/SiCp metal-matrix composites fabricated using the rheocasting technique," Journal of Materials Processing Technology, vol. 80–90, pp. 591–596, 1999.

7. C. García-Cordovilla, J. Narciso, and E. Louis, "Abrasive wear resistance of aluminium alloy/ceramic particulate composites," Wear, vol. 192, no. 1-2, pp. 170–177, 1996.

8. W. Aiguo and H. J. Rack, "Abrasive wear of silicon carbide particulate- and whisker-reinforced 7091 aluminum matrix composites," Wear, vol. 146, no. 2, pp. 337–348, 1991.

9. P. Rodrigo, M. Campo, B. Torres, M. D. Escalera, E. Otero, and J. Rams, "Microstructure and wear resistance of Al–SiC composites coatings on ZE41 magnesium alloy," Applied Surface Science, vol. 255, no. 22, pp. 9174–9181, 2009.

10. Y. Sahin and V. Kilicli, "Abrasive wear behaviour of SiCp/Al alloy composite in comparison with ausferritic ductile iron," Wear, vol. 271, no. 11-12, pp. 2766–2774, 2011.

11. J. K. M. Kwok and S. C. Lim, "High-speed tribological properties of some Al/SiCp composites. I. Frictional and wear-rate characteristics," Composites Science and Technology, vol. 59, no. 1, pp. 55–63, 1999.

12. A. Mazahery and M. O. Shabani, "Existence of good bonding between coated B4C reinforcement and al matrix via semisolid techniques: enhancement of wear resistance and mechanical properties," Tribology Transactions, vol. 56, no. 3, pp. 342–348, 2013.

13. M. O. Shabani and A. Mazahery, "Application of finite element model and artificial neural network in characterization of al matrix nanocomposites using various training algorithms," Metallurgical and Materials Transactions A, vol. 43, no. 6, pp. 2158–2165, 2012.

14. Z. F. Zhang, L. C. Zhang, and Y.-W. Mai, "Particle effects on friction and wear of aluminium matrix composites," Journal of Materials Science, vol. 30, no. 23, pp. 5999–6004, 1995.

15. S. Wilson and A. T. Alpas, "Wear mechanism maps for metal matrix composites," Wear, vol. 212, no. 1, pp. 41–49, 1997.

16. F. Rana and D. M. Stefanescu, "Friction properties of Al-1.5% Mg/SiC particulate metal-matrix composites," Metallurgical and Materials Transactions A, vol. 20, pp. 1564–1566, 1989.

17. S. Kumar and V. Balasubramanian, "Effect of reinforcement size and volume fraction on the abrasive wear behaviour of AA7075 Al/SiCp P/M composites-a statistical analysis," Tribology International, vol. 43, no. 1-2, pp. 414–422, 2010.

18. A. Mazahery and M. O. Shabani, "Mechanical properties of squeeze-cast A356 composites reinforced with B_4C particulates," Journal of Materials Engineering and Performance, vol. 21, no. 2, pp. 247–252, 2012.

19. I. M. Hutchings, "Tribological properties of metal matrix composites," Materials Science and Technology, vol. 10, no. 6, pp. 513–517, 1994.

20. T. Ma, H. Yamaura, D. A. Koss, and R. C. Voigt, "Dry sliding wear behavior of cast SiC-reinforced Al MMCs," Materials Science and Engineering A, vol. 360, no. 1-2, pp. 116–125, 2003.

21. K. R. Brown, M. S. Venie, and R. A. Woods, "The increasing use of aluminum in automotive applications," JOM, vol. 47, no. 7, pp. 20–23, 1995.

6

PHYSICS BEHIND THE OHMIC NATURE IN SILICON CARBIDE CONTACTS

Zhongchang Wang[1]

[1] WPI Research Center, Advanced Institute for Materials Research, Tohoku University, Japan

1. INTRODUCTION

One of the most active fields in semiconductor research is the development of electronic devices capable of function at high powder and high frequency levels, high temperatures, and caustic circumstances. This surge of activity is strongly driven by the urgent desire for replacing the current Si- and GaAs-based electronics because they are unable to operate properly under harsh environmental conditions. As a promising substitute, the wide-band-gap semiconductor, silicon carbide (SiC), has captured considerable attention recently due to its excellent intrinsic properties, which involve large breakdown electric field, high electron saturation drift velocity, strong hardness, and good thermal conductivity. On the other hand, current significant improvements in the epitaxial and bulk crystal growth of SiC have paved the way for fabricating its electronic devices, which stimulates further interest in developing device processing techniques so as to take full advantage of its superior inherent properties.

One of the most critical issues currently limiting the device processing is the manufacturing of reliable and low-resistance Ohmic contacts especially contacts to *p*-type 4H-SiC [1]. The Ohmic contacts are primarily important in SiC devices because a Schottky barrier of high energy is inclined to form at an interface between metal and wide-bandgap semiconductor, which consequently results in low current driving, slow switching speed, and increased power dissipation. Much of effort expended to date to realize the Ohmic contact has mainly focused on two techniques. One is the high-dose ion implantation approach [2], which can increase carrier density in SiC noticeably while reducing its depletion width significantly so that increasing tunneling current is able to flow across the barrier region. The key problem of this method is the easy formation of lattice defects or amorphization during the ion implantation. These defects are unfortunately very stable and need to be recovered via annealing at an extremely high temperature of about 2000 K, thereby complicating the mass production of SiC devices.

The other alternative is to generate an intermediate semiconductor layer with narrower band gap or higher carrier density at the contacts/SiC interface via depositing and annealing technique [3]. To form such layers, a wide range of materials have been examined in a trial-and-error designing fashion, including metals, silicides, carbides, nitrides, and graphite. Of all these materials, the metallic alloys have been investigated extensively, largely because their fabrication process is simple, standard, and requires no exotic materials. In particular, most of research activities have been focused on TiAl-based alloys, the only currently available materials that yield significantly low contact resistance (Ohmic contact) to *p*-SiC [4]. Moreover, they demonstrate high thermal stability. Although a lot of intriguing results have been obtained regarding the TiAl-based contact systems, the mechanism whereby the Schottky becomes Ohmic after annealing has not been well clarified yet. In other words, the key factor to understanding the formation origin of Ohmic contact remains controversial. Mohney et al. [5] proposed that a high density of surface pits and spikes underneath the contacts contributes to the formation of Ohmic behavior based on their observations using scanning electron microscopy and atomic force microscopy. Nakatsuka et al. [6], however, concluded that the Al concentration in the TiAl alloy is important for the contact formation. Using the liquid etch and ion milling techniques, John and Capano [7] ruled out these possibilities and claimed that what matters in realizing the Ohmic character is the generation of carbides, Ti_3SiC_2 and Al_4C_3, between the metals and semiconductor. This, however, differs, to some extent, from the X-ray diffraction (XRD) findings of Chang et al. [8] showing that the compounds formed at the metal/SiC interface are silicides, $TiSi_2$, TiSi, and Ti_3SiC_2. In addition, Ohyanagi et al. [9] argued that carbon exists at the contacts/SiC interface and might play a crucial role in lowering Schottky barrier. These are just a few representative examples illustrating the obvious discrepancies in clarifying the formation mechanism of the Ohmic contact. Taking the amount of speculations on the mechanism and the increasing needs for better device design and performance control, understanding the underlying formation origin is timely and relevant.

To develop an understanding of the origin in such a complex system, it is important to focus first on microstructure characterization. Transmission electron microscopy (TEM) studies by Tsukimoto et al. [10] have provided useful information in this aspect. They have found that the majority of compounds generated on the surface of 4H-SiC substrate after annealing consist of Ti_3SiC_2 and hence proposed that the SiC/Ti_3SiC_2 interface is responsible for the lowering of Schottky barrier in the TiAl-based contact system. However, the role of this interface in realizing the Ohmic nature is still unclear. It is not even clear how the two materials atomically bond together from their experiments, which is very important because it may strongly affect physical properties of the system. Theoretically, we have calculated the atomic structures, adhesive energies, and bonding nature of the SiC/Ti_3SiC_2 interface [11]. However, this calculation does not actually reveal the formation mechanism of Ohmic contact because it only addresses the interface structure. Furthermore, lacking essential experimental information about the interfacial atomic-scale structure, such calculations have been incomplete.

Recent advances in the high-angle annular-dark-field (HAADF) microscopy [12,13], the highest resolution, have enabled atomic-scale imaging of a buried interface. However, direct interpretation of the observed HAADF images is not always straightforward because there might be abrupt structural discontinuity, mixing of several species of elements on individual atomic columns, or missing contrasts of light elements. One possible way out to complement the microscopic data is through atomistic calculation, especially the first-principles calculation. As well known, the atomistic first-principles simulations have long been confirmed to be able to suggest plausible structures, elucidate the reason behind the observed images, and even provide a *quantitative* insight into how interface governs properties of materials [14,15]. Consequently, a combination of the state-of-the-art microcopy and accurate atomistic modeling [16] is an important advance for determining interface atomic-scale structure and relating it to device properties, revealing, in this way, physics origin of the contact issues in SiC electronics.

In addition to determining atomic structure of the 4H-SiC/Ti_3SiC_2 interface, the goal of this work is to clarify the formation mechanism of the TiAl-based Ohmic contacts so as to provide suggestions for further improvement of the contacts. 4H-SiC will hereafter be referred to as SiC. In this Chapter, we will first attribute qualitatively the formation of ohmic contacts in the TiAl-deposited SiC system to an epitaxial and atomically abrupt interface between the SiC and Ti_3SiC_2 generated *via* annealing. The interface presumably serves as a primary current-transport pathway to lower the Schottky barrier formed at the interface between the deposited metals and SiC. Further quantitative studies reveal that the barrier mitigation arises from trapping of an atomic layer of carbon at the SiC/Ti_3SiC_2 interface, which assists the electron transport across the SiC [17,18]. The considerations on the role played by interface do not, however, exclude another possibility that the Ti_3SiC_2 atomic layers can be generated inside the SiC bulk interiors, which presents a behavior that may differ from that of their bulk [19]. Combining the state-of-the-art TEM with atomistic first-principles calculations, we demonstrate the presence of an atomic-scale Ti_3SiC_2-like bilayer embedded in the SiC, forming an atomically ordered multilayer that exhibits an unexpected electronic state with point Fermi surface. The valence charge is confined to a large extent to within the bilayer in a spatially connected manner, serving possibly as a conducting channel to enhance the current flow over the semiconductor. Further investigation into the contact regions unveils another new opportunity to allow the electron transport across the semiconductor, namely, *via* the terraces formed at the SiC/Ti_3SiC_2 interface [20]. Experimentally, the formed carbide Ti_3SiC_2 is demonstrated to bond directly to the silicon carbide at the terraces in an epitaxial and atomically ordered fashion, regardless of dimension of the terraces. There appear pronounced gap states at Fermi level in the semiconductor layers around the terraces, and charges are accumulated heavily around the terraces in a connected and broadly distributed manner. The presence of the metallicity and the likelihood to act as electron conduction channels to enable the current flow over the semiconductor make the terraces at the interface one of the origins underlying the ohmic contact in silicon carbide electronics. We therefore demonstrate in this chapter that origin of the long-standing contact issue in SiC devices can be understood and technologically manipulated at the atomic level, and suggest the key physical factors for establishing the ohmic nature.

2. ROLE OF COHERENT SIC/TI_3SIC_2 INTERFACE

The *p*-type 4H-SiC epitaxial layers (5-μm thick) doped with aluminum (N_A = 4.5 × 10^{18} cm^{-3}) which were grown on undoped 4H-SiC wafers by chemical vapor deposition (manufactured by Cree Research, Inc.) were used as substrates. The 4H-SiC substrates had 8°-off Si-terminated (0001) surfaces inclined toward a [¯2110] direction because only 4H-type structure of SiC with polymorph (e.g. 3C, 4H, 6H, 15R etc.) was controllable by lateral growth of the epitaxial layers parallel to (0001)-oriented surface. After chemical cleaning of the substrate surface, a 10 nm-thick sacrificial oxide (SiO_x) layer was grown on the SiC substrate by dry-oxidation at 1423 K for 60 min. The electrode patterns were made by removing the SiO_x layers, where contact metals were deposited by dipping in 5 % diluted hydrofluoric acid solution for 1 min using a photolithography technique. Prior to the deposition of contact materials, the substrates were cleaned by deionized water. Then, Ti and Al stacking layers with high purities were deposited sequentially on the substrate in a high vacuum chamber where the base pressure was below 5 × 10^{-6} Pa. The thicknesses of the Ti and Al layers investigated in this study are 100 nm and 380 nm, respectively, and these layer thicknesses were chosen to give the average composition of the Ti(20 at%) and Al(80 at%), where the layer thicknesses were measured by a quartz oscillator during deposition. The reasons to choose this average composition was that aluminum rich (more than 75 at%) in TiAl contacts were empirically found to be essential to yield low contact resistance, resulting from formation of the Ti_3SiC_2 compound layers. After depositing, the binary TiAl contact layers were annealed at 1273 K for a storage time of 2 min in an ultra-high vacuum chamber where the vacuum pressure was below 1×10^{-7} Pa.

The surface morphology of the TiAl contact layers on 4H-SiC after annealing was observed using a JEOL JSM-6060 scanning electron microscope (SEM). Microstructural analysis and identification of the Ti_3SiC_2 layers at the contact layers/4H-SiC interfaces after annealing was performed using X-ray diffraction (XRD) and cross-sectional TEM. For XRD analysis, Rigaku RINT-2500 with Cu *Kα*radiation operated at 30 kV and 100 mA was used. In particular, the interfacial structures and an orientation relationship between the contact layers and the 4H-SiC substrates were characterized by cross-sectional high-resolution TEM observations and selected area diffraction pattern (SADP) analysis, respectively, using a JEOL JEM-4000EX electron microscope operated at an accelerating voltage of 400 kV, where the point-to-point resolution of this microscope was approximately 0.17 nm. Z-contrast images were obtained using a spherical aberration (C_s) corrected scanning transmission electron microscope (STEM) (JEOL 2100F), which provides an unprecedented opportunity to investigated atomic-scale structure with a sub-Å electron probe. Thin foil specimens for the TEM and STEM observations were prepared by the standard procedures: cutting, gluing, mechanical grinding, dimple polishing, and argon ion sputter thinning techniques.

Calculations of electronic structure and total energy were carried out using the Vienna *ab initio*simulation package (VASP) within the framework of density functional theory (DFT) [21]. The projector augmented wave method was used for electron-ion interactions, and the generalized gradient approximation of Perdew et al. (PW91) was employed to describe the exchange-correlation functional. The single-particle Kohn-Sham wave function was expanded using plane waves with different cutoff energies depending on the calculated systems. Sampling of irreducible Brillion zone was performed with a regular Monkhorst-Pack grid of special *k* points, and electronic occupancies were determined according to the Methfessel-Paxton scheme. Independent *k* point convergence tests were conducted for distinct supercells. Ground state charge densities were calculated self-consistently using a Pulay-like mixing scheme and the stable blocked Davidson minimization algorithm. Total energies were calculated using the linear tetrahedron method with Blöchl corrections, which eliminates broadening-related uncertainties. All atoms were fully relaxed using the conjugate gradient algorithm until the magnitude of the Hellmann-Feynman force on each atom converged to less than 0.05 eV/Å, yielding optimized structures.

To determine the most stable interface theoretically, one first has to establish feasible models on the basis of the distinct terminations and contact sites and then compare them. However, a direct comparison of total energies of such candidate models is not physically meaningful since interfaces might have a different number of atoms. On the other hand, the adhesion energy (W_{ad}), which is key to predicting the mechanical properties of an interface, is physically comparable. Generally, the W_{ad} is defined as the energy required to reversibly separate an interface into two free surfaces, neglecting plastic and diffusion degrees of freedom. The energy needed in actual cleavage experiments is always greater than the W_{ad} owing to plastic deformation, but the extent of plastic deformation relies on theW_{ad}. Formally, the W_{ad} can be expressed in terms of surface and interfacial energies or by the difference in total energy between the interface and isolated slabs [22,23]:

$$W_{ad} \equiv \sigma_1 + \sigma_2 - \sigma_{12} = (E_1 + E_2 - E_{IF}) / A \tag{1}$$

Here σ_i is the surface energy of slab *i*, σ_{12} is the interface energy, E_i is the total energy of isolated slab *i*, E_{IF} is the total energy of the interface system, and *A* is the total interface area. In general, two steps can be taken to estimate the W_{ad}. First, the total energies were calculated for a series of separations as two rigid slabs were brought increasingly closer from a large initial separation. As a consequence, the calculated total energies were found to behave like a parabola, passing through a minimum at the equilibrium separation. The unrelaxed W_{ad} was obtained by computing the energy difference between the interface at the equilibrium state and the unrelaxed isolated slab. Next, each isolated slab as well as interfacial slab was allowed to optimize fully, yielding an estimation of relaxed W_{ad}.

2.1. Atomic-Scale Structures Of The Ohmic Contacts

The electric properties for the TiAl contact systems before and after annealing are meaured first to verify the formation of Ohmic contact. The current almost keeps zero before the annealing, while increasing nearly linear with the rise of applied bias, which unambiguously confirms the formation of the Ohmic contacts after annealing. Further XRD analyses demonstrate that a new reaction product, the ternary Ti_3SiC_2, is generated after annealing, which shows a strongly (0001)-oriented texture. The SiC retains (0001)-oriented texture after annealing, thereby facilitating development of hetero-epitaxy between reaction products and substrates. The TEM imaging reveals that no any other compounds contact directly the SiC surface, thereby ensuring an exclusive contact of Ti_3SiC_2 to SiC. Since the carbide itself is metallic in nature, the lowering in Schottky barrier in the TiAl-based contacts is hence attributed qualitatively to the epitaxial and atomically sharp SiC/Ti_3SiC_2 interface. A careful indexing of the selected area diffraction pattern (SADP) at the contacts/SiC interface reveals that the formed Ti_3SiC_2 layers have epitaxial orientation relationships, that is, (0001)Ti_3SiC_2//(0001)SiC and [0¯110]Ti_3SiC_2//[0¯110]SiC, with the SiC substrate. These orientation relationships are believed to be beneficial for forming a coherent and well matched interface between SiC and Ti_3SiC_2, since they both belong to the hexagonal space group with lattice constants of a = 3.081 Å and c = 10.085 Å for the SiC and a = 3.068 Å and c = 17.669 Å for the Ti_3SiC_2.

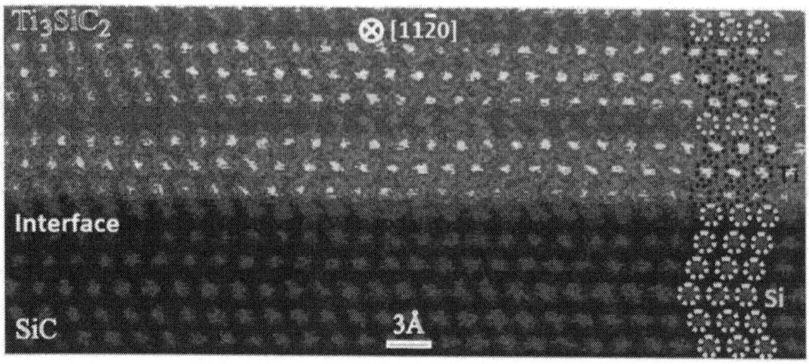

Figure 1. A typical HAADF-STEM image of the SiC/Ti_3SiC_2 interface in the annealed TiAl contact system observed from [11-20] direction. The points at which the phase contrast is no longer periodic in either the Ti_3SiC_2 or SiC define the interfacial region [23].

A representative HAADF image of the SiC/Ti_3SiC_2 interface is shown in Fig. 1, which confirms a clean and atomically sharp contact between the two materials. Since intensity of an atomic column in the Scanning TEM, to good approximation, is directly proportional to the square of atomic number (Z) [24], brighter spots in image represent atomic columns of Ti, while the comparatively darker ones are Si. Not surprisingly, the columns of C are not scattered strongly enough to be visualized owing to its small Z, thereby making this image incomplete. It should be noted that obtaining a signal of pure interfacial carbon is technically very difficult because the specimen can be easily affected by the environmental carbon, thereby precluding the element-selective imaging of carbon. We therefore rely on the first-principles calculations instead to discuss the possibility in the presence of C at the interface, as will be described later. To see the interface clearer, we magnify the cross-sectional HAADF image in Fig. 2(a) and further filter it to reduce noise, as shown in Fig. 2(b). The Si-terminated Ti_3SiC_2 is observed intuitively to make a direct contact with the Si-terminated SiC substrate with interfacial Si atoms of Ti_3SiC_2 sitting above hollow sites of interfacial Si plane of SiC. However, this straightforward interpretation is premature, as will be described later. Since there are no pits, spikes, or dislocations that might act as pathways for current transport, we conclude that this clean and coherent SiC/Ti_3SiC_2 interface should be critical for the formation of Ohmic contact.

To clarify the mechanism, it is prerequisite to determine the atomic structure of the SiC/Ti_3SiC_2interface via complementing the obtained HAADF image (Fig. 1). We have considered a total of 96 candidate interfacial geometries using bulklike slabs, taking into account termination effect, stacking sequence, and full optimization. From the bulk 4H-SiC and Ti_3SiC_2 structures and the relative stacking order of Ti and Si, the observed image in Fig. 2(a) can be intuitively fitted by a SiSi model shown inFig. 2(c). In this model, the interfacial Si atoms of Ti_3SiC_2 sit above the hollow sites of interfacial Si plane of SiC, where the optimal distance between interfacial Si-Si planes (denoted as *d*1 in Fig. 2(b)) and that between interfacial Si-Si atoms projected onto paper plane (denoted as *d*2 in Fig. 2(c)) are calculated to be 2.13 and 2.53 Å, respectively. These distances, however, deviate severely from their average experimental values, 2.5 Å and 2.8 Å, which are obtained by characterizing quantitatively the HAADF image (Fig. 2(a)). In addition, an examination of interface stability by calculating the W_{ad}indicates that this SiSi model is not favored (1.62 J/m^2). It is even less stable than the model with interfacial Si of Ti_3SiC_2 resting straight atop the interfacial Si of SiC (2.58 J/m^2), which contravenes again the observed image.

To resolve these paradoxes, we notice that a possibility might be ignored, that is, the unseen C might be trapped at the interface, altering local environment there. To test this scenario, we established a new model (named SiCSi) by introducing C into the interfacial layer from the consideration of crystal extension and stacking sequences. The calculated W_{ad} of this SiCSi model is 6.81 J/m^2, which indicates that interface is indeed strengthened substantially after incorporating C. Further examination of the relaxed atomic geometry (Fig. 2(d)) reveals that the incorporation of C does not induce a significant structural transformation. Namely, the two Si layers proximal to the interface maintain the stacking seen in Fig. 2(c), thus matching the HAADF image geometrically. Quantitatively, the *d*1 and*d*2 distances are now 2.53 and 2.81 Å, respectively, very close to the experimental values. Therefore, the introduction of interfacial C monolayer resolves the inconsistencies between simulations and experiments.

Figure 2. a) Magnified HAADF image of the SiC/Ti_3SiC_2 interface. (b) The same image as in (a) but has been low-pass filtered to reduce noise. Relaxed SiC(0001)/Ti_3SiC_2(0001) interface models (c) without (SiSi) and (d) with (SiCSi) interfacial C atoms. The distance between interfacial Si-Si layers is represented by *d*1 and that between interfacial Si-Si atoms projected onto the paper plane by *d*2. The interfaces are represented by an arrow [23].

2.2. Electronic Structure and Bonding

Calculations of *p*-type Schottky barrier height (SBH) reveal that the interface with C (0.60 eV) has much lower SBH than the interface without C (1.05 eV), suggesting that the trapped C assist the lowering of SBH. To shed light on origin of the decrease in SBH and junction strengthening in the SiCSi interface, we characterized thoroughly interfacial electronic states and bonding nature. Figure 3(a) shows a planar-averaged charge-density difference along interface normal, where there appears a more dramatic accumulation of charge within the interfacial region for the interface with C. This indicates that the covalent bonding is strengthened in the SiCSi case. In addition, we note that the planar-averaged density difference for the SiCSi more prominently deviates from zero around interface, reflecting more significant charge transfer between the SiC and Ti_3SiC_2 slabs. Moreover, charge is observed to be depleted noticeably in both the sub-interfacial SiC and Ti_3SiC_2 region for the SiCSi, suggesting that the atoms second nearest to the interface contribute to interfacial bonding. These missing charges, to a large extent, make their way onto the more electronegative C ions, indicative of the formation of ionic bonding.

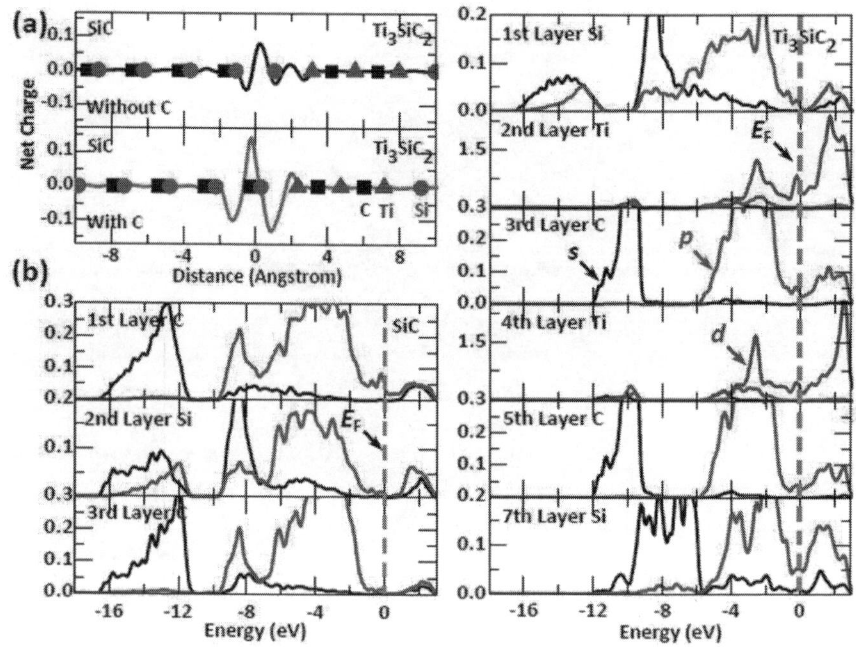

Figure 3. a) Planar-averaged charge-density difference for the relaxed interfaces with and without C along [0001] direction. (b) Density of states projected onto the atomic layers close to the relaxed SiCSi interface. The left bottom panel shows the PDOSs of SiC layers and the right one those of Ti_3SiC_2layers. The first layer is the atomic layer proximal to interface. The *E F*is set to zero [23].

We then presented in Fig. 3(b) DOS projected on selected atomic layers of the SiCSi interface. A key feature in this figure is that a strong interaction is observed between the sub-interfacial Ti *d* and Si *p*states below Fermi level (*E F*), which continues well into SiC surface, inducing noticeable gap states in the interfacial C at *E F* . This means that the interfacial C layer is metalized, indicative of possible electric conductivity. In fact, the gap states can extend as far as they can into deeper layers of SiC, as there appear weak but visible peaks at *E F*in the

PDOSs of the 2nd and 3rd layers of SiC. Therefore, a local weak metallicity might occur at top few layers of the semiconductor surface, which could enable current flow through the SiC. We also note significant hybridization between the interfacial C *sp* and Si *sp* states, suggestive of covalent bonding at interface.

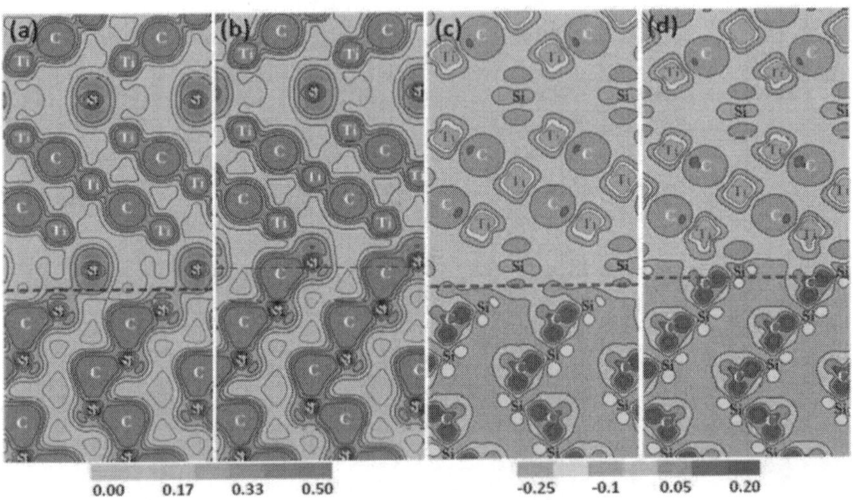

Figure 4. Contour plots of charge densities for (a) SiSi and (b) SiCSi interfaces taken along (11-20) plane. The interface is represented by a horizontal line and the atoms that intersect the contour plane are labeled. Corresponding contour plots of charge-density differences for (c) SiSi and (d) SiCSi interfaces [23].

Figure 4 shows contour plots of charge densities and their differences along (11-20) plane for the optimized SiSi and SiCSi interfaces. We notice in Fig. 4(b) that the bonding interaction between interfacial Si and C for the SiCSi interface is remarkably similar to the Si-C interaction deeper into SiC: the majority of charge is localized on C with humps directed towards their neighboring Si. We thus conclude that the interfacial bonding for the SiCSi is of mixed covalent-ionic nature. The interfacial bonds for the SiSi interface, however, have covalent nature with a small amount of charge accumulated within the interfacial region (Fig. 4(a)). In addition, the amount of charge accumulated on the interfacial Si-Si bonds of SiSi (Fig. 4(c)) is far less significant than that on the interfacial Si-C bonds of SiCSi (Fig. 4(d)). This heaver charge accumulation in the case of SiCSi, together with its mixed covalent-ionic character at interface, accounts for the largest W_{ad} associated with the SiCSi interface.

2.3. Quantum Electron Transport Properties

Although the charge-distribution analysis can reveal valuable information on interfacial bonding, it provides restrained insight into how electrons distribute around E_F, which matters because density around E_F directly determines the current transmission. Figure 5 illustrates an electron-density isosurface and its slice along the (11-20) plane for the optimized SiCSi interface around E_F. From Fig. 5(a), one can see that charges surrounding interfacial Si are connected and broadly distributed in a sheet-like fashion, which suggests possible electrical conductivity through this region. In addition, there also appear heavily accumulated electrons within the interfacial area, which are connected along the interface and extended as far as several atomic layers into the SiC. These characters can also be confirmed from the slice plot in Fig. 5(b), meaning that current might flow over top few atomic layers of semiconductor, thereby causing Ohmic property. As expected, the electron density at E_F is extremely high for the Ti_3SiC_2 (*i.e.*, sea of electrons) but becomes nil for the SiC layers away from interface (Fig. 5(b)), which can be understood from their intrinsic metallic and semiconducting nature [25].

To examine electrical conductivity and gain insight into how the interface influences current transport, we devised a two-probe system [26], Ti/Ti_3SiC_2/SiC/Ti_3SiC_2/Ti, and investigated nonequilibrium quantum transport properties. Figure 6 schematically shows a model of the sandwich transport system, which can be divided into a left semi-infinite electrode, a scattering region, and a right semi-infinite electrode. The atomic and electronic structures of the semi-infinite Ti electrodes are assumed to be the same as those of bulk Ti. On the other hand, the electronic states of scattering region are calculated self-consistently. The scattering region consists of hexagonal SiC and Ti_3SiC_2 layers and the periodic boundary conditions are imposed along the directions parallel to the interface. The SiC/Ti_3SiC_2interface could be either the SiSi or SiCSi, whereas other interfaces are maintained identical for the sandwich systems. In this sense, the difference between the two systems can be mainly attributed to their differing SiC/Ti_3SiC_2

interfaces. Furthermore, we also calculated the Ti/SiC/Ti system, wherein the SiTi model was taken as the SiC/Ti interface.

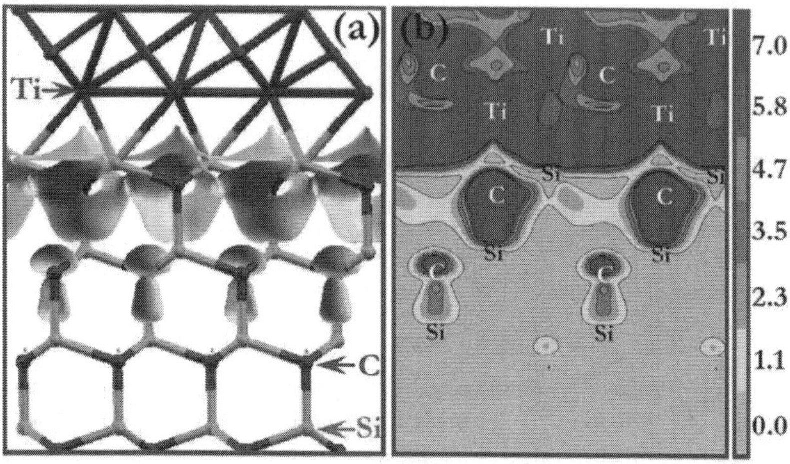

Figure 5. a) Isosurface and (b) electron density plot along the (11-20) plane in the energy window (*E F*-0.5 eV,*E F*) for the SiCSi interface. The interface is marked by two arrows [18].

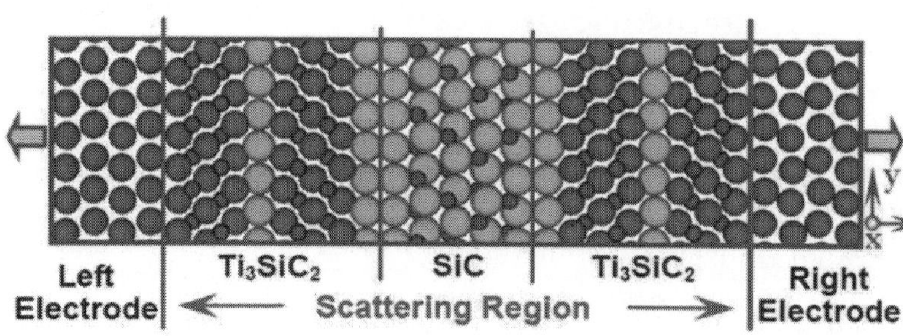

Figure 6. Schematic illustration of a two-probe $Ti/Ti_3SiC_2/SiC/Ti_3SiC_2/Ti$ quantum transport system. The system has infinite extent in the (x, y) direction and extends to $\pm\infty$ in the z direction. The SiC/Ti_3SiC_2 interfaces shown in Fig. 2 are adopted [18].

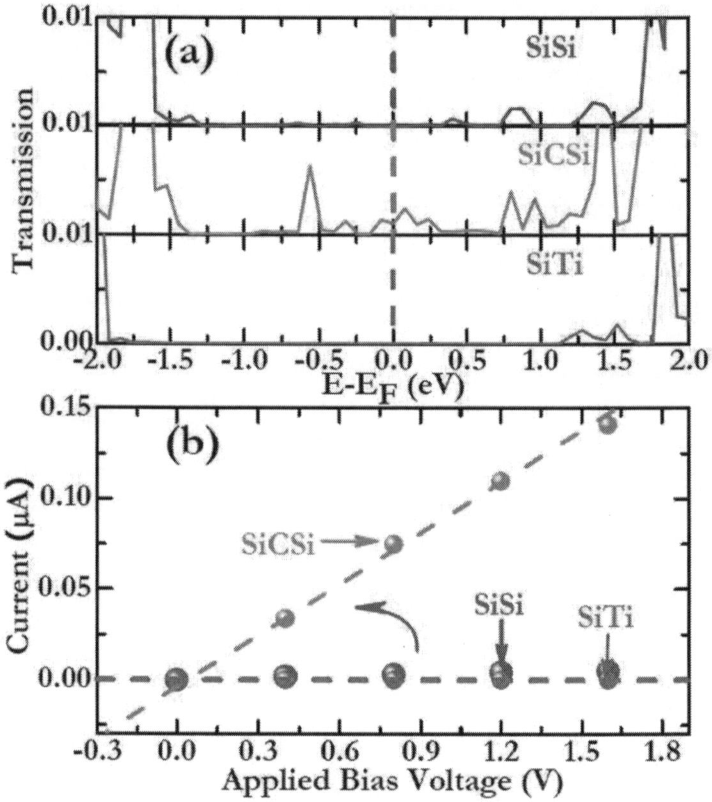

Figure 7. a) Transmission spectra under 0 V and (b) current-voltage characteristics for the sandwich systems involving the interfaces containing direct Si-Si bonding (SiSi), Si-C-Si bonding (SiCSi), and the direct contact of Ti to SiC (SiTi). Refer to Fig. 2 for their corresponding interfacial configurations [18].

Figure 7(a) shows transmission spectra for the relaxed SiSi, SiCSi, and SiTi systems, where one can see that the spectra differ from one another suggesting variations in electronic structures with interface geometries. The most interesting feature is the presence of transmission peaks at *E F* for the SiCSi, which is attributable to the electrons distributed around the interface at *E F* . Further calculations on electrical properties (*e.g., I–V* curve) for the three systems reveal that the current in the SiCSi case increases much faster than either the SiSi or SiTi case as the applied bias voltage increases (Fig. 7(b)), which can be explained by its lowest SBH.

We further examined how applied bias voltages vary from the interface to the SiC region by analyzing the difference in effective potential along the (11-20) plane between the bias voltage of 0.4 V and the one of 0.0 V for the relaxed system. The voltage is found to drop less intensively from the interface to the SiC in the SiCSi case, suggestive of less Schottky nature. Finally, in comparing the general trend of the calculated *I–V* with that of our experimental curve, we find that they agree qualitatively: both the annealed specimen and the SiCSi model clearly show Ohmic behavior, thereby validating the application of the SiCSi model to describe the Ohmic contacts in the TiAl-based system. In addition to the role of this interface, we also found that an atomic-scale Ti_3SiC_2-like bilayer can be embedded in the SiC interior, forming an atomically ordered multilayer that exhibits an unexpected electronic state with point Fermi surface. The valence charge is confined largely to this bilayer in a spatially connected fashion, serving possibly as a conducting channel to enhance the current flow over the semiconductor.

3. ATOMIC-SCALE TI_3SIC_2 BILAYERS EMBEDDED IN SIC

3.1. Atomic Structure Of The Embedded System

Figure 8(a) shows a HAADF image of the annealed TiAl/SiC system, where the SiC substrate is covered entirely by a layered compound, Ti_3SiC_2, as reported previously. In addition to the formation of this epitaxial and coherent SiC/Ti_3SiC_2 interface, another interesting feature is that an atomic-scale bilayer is generated in the SiC interior (marked by a square in Fig. 8(a)), which is located approximately 9.5 nm away from the interface. An enlarged image of the region surrounding the bilayer shows that it has a Ti_3SiC_2-like structure, as shown in Fig. 8(b), where brighter spots represent atomic columns of Ti (smaller circles), while dark ones those of Si (larger circles), since intensity of an atomic column in STEM is, to a good approximation, directly proportional to $Z^{1.7}$ (Z: atomic number) [24].

Evidently, the bilayer is embedded in the SiC in an atomically coherent and ordered fashion with no transitional or intermixing layers (see Fig. 8(c)). In view of bulk structures of 4H-SiC and Ti_3SiC_2 and the relative stacking sequence of Ti and Si, the image in Fig. 8(c) can be qualitatively fitted by an energetically stable model shown in Fig. 8(d). In this model, the optimal distances between layers around the bilayer (denoted as *L1* to *L5* in Fig. 8(d)) are calculated to be 2.52, 2.27, 2.28, 2.40, and 2.57 Å, respectively, very close to experimental values of 2.5, 2.2, 2.3, 2.3, and 2.7 Å estimated via quantitative characterization and averaging of different sites in the HAADF images. In addition, the calculated distances between neighboring atoms (labeled *d1* to *d5* in Fig. 8(d)) (2.67, 2.44, 2.44, 2.55, and 2.72 Å) also approach the experimental values (2.7, 2.3, 2.4, 2.6, and 2.7 Å). These mean that the model constructed (Fig. 8(d)) matches the HAADF image (Fig. 8(c)) both qualitatively and quantitatively in light of energetics and atomic distances.

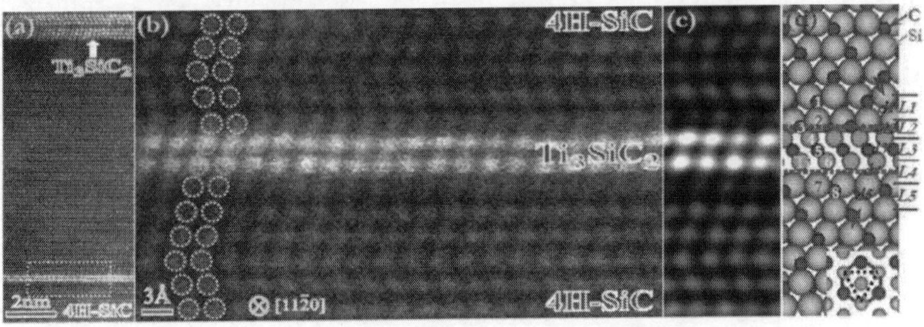

Figure 8. a) HAADF-STEM image for the Ti_3SiC_2-like bilayer embedded in 4H-SiC in the annealed TiAl contact system viewed along [11-20] direction. The bilayer sits about 9.5 nm below the SiC/Ti_3SiC_2interface. (b) Magnified HAADF image of the region marked in (a) by a dotted square. An overlay is given, where the big circles indicate Si and small ones Ti. (c) The same image as in (b) but has been low-pass filtered to reduce noise. (d) The optimized SiC/Ti_3SiC_2/SiC multilayer model. The distances between layers around the bilayer are represented by *Lm* (*m* = 1 to 5) and those between neighboring atoms projected onto the paper plane by *dm*. The atomic layers are labeled 1 through 8. The inset shows top view of arrangement of Ti and its neighboring Si and C, where the Ti occupies atop the center of the hexagon composed of Si and C [19].

3.2. Formation Of Point Fermi Surface

To gain insight into how the embedded layer influences SiC electronically, we present in Fig. 9 band structure and density of states (DOS) of the multilayer system calculated using the optimal atomic geometry (Fig. 8(d)). Unexpectedly, several bands with a quadratic dispersion cross the Fermi level (E_F) at a single Γ point (Fig. 9(a)), rendering conduction bands (CB) and valence band (VB) touch at their tips and hence the multilayer become a gapless semiconductor. Inspection throughout the Brillion zone verifies that the Fermi surface crossing is a single point, which determines E_F, as in the bilayer graphene with no external stimuli (bands in graphene extend linearly both to lower and higher energy from point Fermi surfaces, as referred to as "massless Dirac"). This crossing of bands is confirmed in the DOS showing a curious vanishing of states at E_F for both spins (Fig. 9(b)) and further verified in a surface plot of the two bands proximal to E_F in a small k space presenting gapless character at the Γ point and a gap in the region away from the crossing point (Fig. 2(c)).

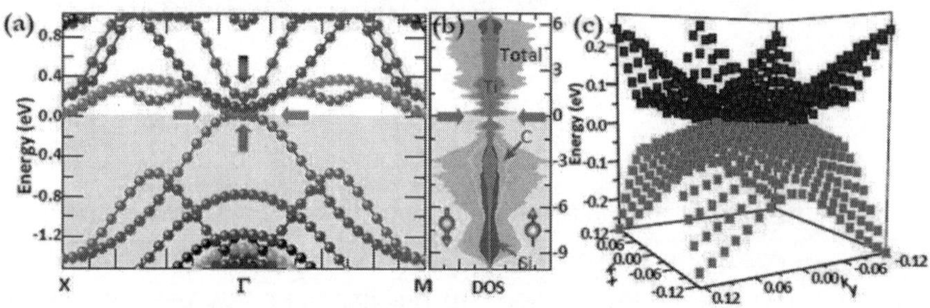

Figure 9. a) Blowup of band structure around E_F shown on the *xy* plane with $X = \pi/a(1, \bar{1}\bar{1}, 0)$ and $M = \pi/a(1,1,0)$, where the "*a*" is in-plane lattice constant. Note that the point Fermi surface is at Γ where bands cross precisely at E_F. (b) Total DOS and PDOS plots of C, Si, and Ti atom contributions for the optimized SiC/Ti_3SiC_2/SiC multilayer, showing that bands surrounding E_F have characters of Ti in the bilayer. (c) "Surface" plot of the two bands that cross the E_F, centered surrounding the Γ point. The E_F position is aligned to zero [19].

In addition, extensive calculations using the LDA and PBE functional corroborate once again the peculiar crossing of the bands (band structure and DOS spectra are almost identical to those calculated using the PW91), which therefore indicates that the crossing at E_F is not an accidental degeneracy arising from the applied functional. This behavior is quite unusual, as it turns up neither in the SiC bulk showing E_F lying in a gap between states nor in the Ti_3SiC_2 bulk showing E_F lying in the middle of a band of electronic states, which can be attributed to the structural symmetry with Ti atoms sitting exactly atop center of the hexagon composed of Si and C (corresponding to the Γ point in *k* space), as shown in inset of Fig. 8(d). Further calculations reveal that this gapless nature appears in the multilayer consisting of a Ti_3SiC_2-like monolayer embedded in SiC as well, indicating that the unique state could be understood upon quantum confinement effect and the interface phenomenon accompanied by the polarity discontinuity. However, our additional calculations suggest that the point Fermi surface vanishes when a Ti_3SiC_2 trilayer is hypothetically embedded within SiC, which takes on metallic states with three bands (arise from the three Ti layers) crossing E_F, similar to what is seen in the band structure of Ti_3SiC_2 bulk. This transition from the zero-gap semiconductor to metal therefore highlight the importance of quantum confinement in inducing the point Fermi surface, since additional layer is believed to relieve the confinement effect and produce more states.

3.3. Electronic States

Further investigation of DOS projected on selected atomic layers provides evidence that the bands close to E_F involve dominantly Ti 3*d* states (layers 4 and 6 in Fig. 8(d)). These states undergo a sudden vanishing at E_F (Fig. 10), which is in stark difference from what is seen in the DOS projected (PDOS) on Ti in bulk Ti_3SiC_2 presenting continuous states at and around E_F (lines in Fig. 10). In addition, there are notable electronic states in the forbidden gap of bulk SiC for the C and Si in the multilayer, which is definitely attributable to the embedded bilayer. The presence of the induced gap states, in particular, those below the CB minimum of bulk SiC but close to E_F, can readily present a trap for CB electrons, which could modify electronic behaviors of the originally insulating SiC and thus be relevant for the current flow over semiconductor in the TiAl-deposited SiC system. Moreover, we note that overall features of PDOS for identical atom species may even differ from one another, which can be ascribed to their different bonding environments.

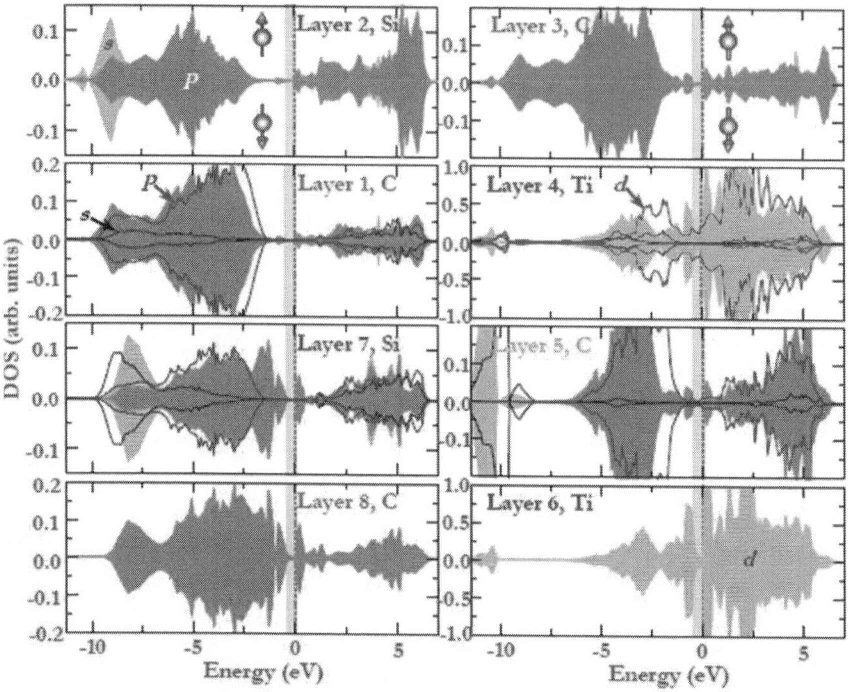

Figure 10. DOS projected on selected atomic layers in the multilayer system. Left panel gives the PDOSs of SiC layers and right panel those of Ti_3SiC_2 layers. The lines show PDOSs of the corresponding atoms in the bulk as a reference. The E_F is set to zero [19].

To shed further light on bonding nature and charge distribution in the multilayer, we present contour plots of charge density (Fig. 11(a)) and its difference (Fig. 11(b)) along the (11-20) plane. From the figures, we notice that (i) majority of charge is localized on C atoms with humps distorted toward their neighboring atoms, suggesting that bonds in both SiC and bilayer are of a mixed covalent-ionic nature, (ii) charge distribution on C in SiC exhibits more pronounced lobes than that on C in the bilayer (Fig. 11(a)), indicative of more covalent element for bonds in SiC, and (iii) ionicity originates from the large charge gain on C at an expense of charge loss on its neighboring cations (Fig. 11(b)). Further examination on electron distribution around E_F reveals that valence electrons are confined, to a large extent, to within the bilayer (Fig. 11(c)), in good agreement with the DOS analysis (Fig. 3). These charges are spatially connected and broadly distributed surrounding the bilayer with a small degree of leakage into as far as two atomic layers of SiC (Fig. 11(c)), as is further confirmed from a slice plot inFig. 11(d). This implies that the bilayer buried in between the SiC may act as a conducting channel so as to enhance current flow over the semiconductor.

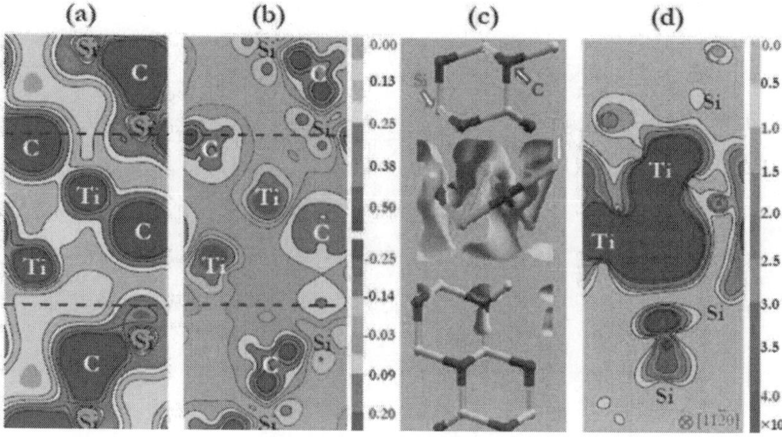

Figure 11. Contour plot of (a) charge density and (b) density difference for the multilayer system viewed along the (11-20) plane. The difference of charge density gives the redistribution of charge in the system relative to its isolated one. The upper scale denotes the magnitude of charge in (a) and the lower scale that of charge difference in (b). (c) Isosurface and (d) electron density plot along the same plane as in (a) in the energy window (E_F-0.5 eV, E_F) [19].

4. TERRACES AT OHMIC CONTACT IN SIC ELECTRONICS

Combining imaging with atomistic simulations, we determine the atomic-scale structures of terraces in between SiC and its contacts and relate the structures to their electronic states and bonding nature, aimed at revealing the impact of the terraces on the contacts of SiC electronics. The terraces were first characterized using the high-resolution TEM (HRTEM) and scanning TEM (STEM), upon which the first-principles calculations were performed. The combined study allows a deeper understanding of the role played by terraces in the ohmic contact formation on a quantum level. The terraces are structurally epitaxial, coherent and atomically ordered, and theoretically predicted to have electronic states at Fermi level (E_F) regardless of their dimension.

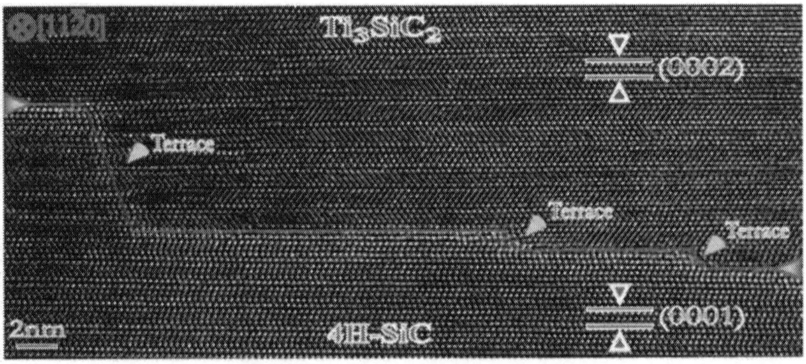

Figure 12. A cross-sectional HRTEM image of the contact of the formed Ti_3SiC_2 to the 4H-SiC substrate viewed from the [11-20] direction. The two materials are demarcated by a zigzag line. Terraces with varying dimensions are observed, as indicated by arrows [20].

4.1. Atomic Structures Of Terraces

Figure 12 shows a typical cross-sectional HRTEM image of the contact of Ti_3SiC_2 to SiC which includes terraces of various dimensions. Well arranged (000*l*)-oriented lattice fringes can be observed in both the SiC and Ti_3SiC_2 layers. The points at which the phase contrast is no longer periodic define the interfacial area, as indicated by a zigzag line. Evidently, the SiC substrate is covered entirely by the Ti_3SiC_2 layers even in the terrace region (arrows), meaning a direct contact of Ti_3SiC_2 to SiC at the terraces at atomic scale. The terraces are atomically abrupt and ordered, showing no amorphous layers, secondary-phase layers, contaminants, or transition areas. A small number of misfit dislocations are identified at the contact region (not shown here) due to the small lattice mismatch of SiC to Ti_3SiC_2(less than 0.5%). Further analyses of selected area diffraction patterns reveal that the Ti_3SiC_2 layers have epitaxial orientation relationships, (0001)Ti_3SiC_2//(0001)SiC and [0-110]Ti_3SiC_2//[0-110]SiC, with the SiC substrate.

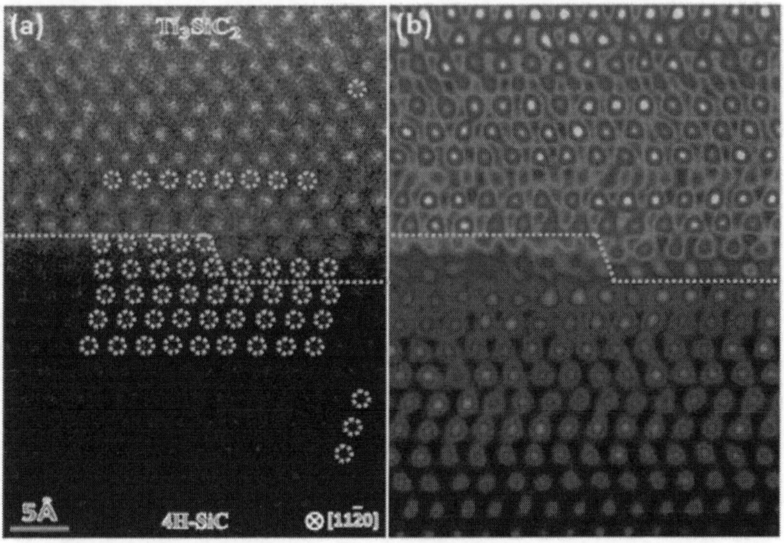

Figure 13. a) A typical HAADF-STEM image of a small terrace observed from the [11-20] direction. The terrace is indicated by a zigzag line. Bigger dotted circles denote Ti and the smaller ones Si. (b) The same image as in (a) but has been filtered to reduce noise [20].

In general, this contact region contains terraces with a wide variety of dimensions that can be affected by numerous factors. However, to develop an understanding of such a complex contact, it is important to first focus on representative terraces. Here, we choose purposely three species of terraces based on the dimension: small, intermediate, and large terrace. The corresponding HAADF images are presented in Figs. 13–15, which confirm the atomically abrupt and ordered terraces. Moreover, heteroepitaxy is retained between the SiC and Ti_3SiC_2 for each terrace. Since the intensity of an atomic column in the HAADF image is proportional to $Z^{1.7}$ (Z: atomic number) [24], brighter spots in the images represent atomic columns of Ti while darker ones are Si. In view of the atomic arrangements in the SiC and Ti_3SiC_2 bulks, we define the terrace by a dashed line and identify the atoms in a few layers close to the line as in the terrace region hereafter. For the small terrace, the image can be intuitively interpreted as a bonding of Ti (Si) in Ti_3SiC_2 to Si in SiC (Fig. 13(b)). However, this is not the case for the intermediate terrace showing Si-Si bonding at the hollow site (Fig. 14(b)). Further interpretation of the HAADF mage of large terrace (Fig. 15(b)) reveals a Si-Si bonding as well. Not surprisingly, atomic columns of carbon are not scattered strongly enough to be visualized due to its small *Z*, rendering the image incomplete. Further complementing of these images so as to relate their atomic structures to their electronic properties requires a combination of imaging with the first-principles modeling.

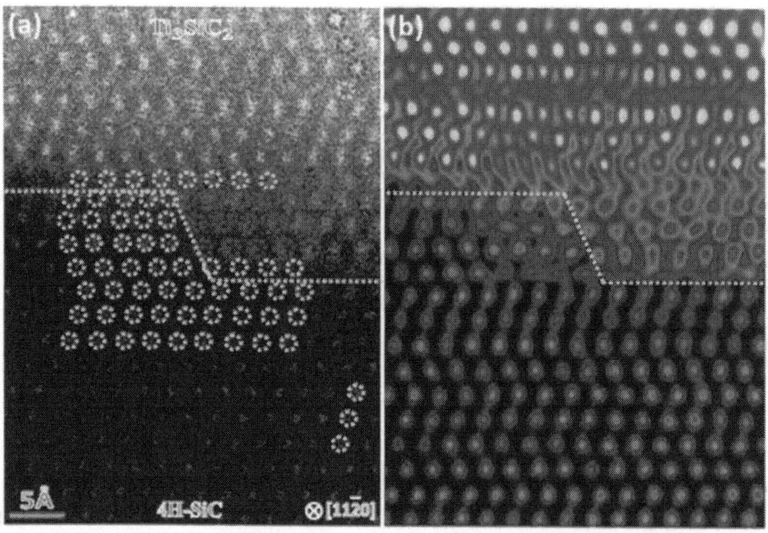

Figure 14. A typical HAADF image of an intermediate terrace observed from the [11-20] direction. The terrace is indicated by a zigzag line. (b) The same image as in (a) but has been filtered to reduce noise [20].

4.2. Electronic States and Bonding Nature Of The Terraces

To gain insight into electronic properties of the terraces and the role they played in the ohmic contact formation, we perform first-principles calculations on the three typical terraces. Upon a consideration of bulk structures of SiC and Ti_3SiC_2, the aforementioned orientation relations, and relative stacking sequence near terraces shown in the HAADF (Figs.

13–15), atomic models of the three representative terraces were established (Fig. 16), taking into account full structural relaxation. It should be noted that these models may not exactly reflect the real terraces because it is the extreme difficult to interpret directly the HAADF images owing to the intricate atomic arrangements around the terraces and to the invisible C atoms. However, these models are constructed upon a careful consideration of the space filling and local chemical environment, and importantly they exhibit the typical variations of the contact with terraces, which should be useful as an initial stage to look at the electronic states of terraces. It is also worthy of mentioning that neither the total nor the interface energies could be applicable to justify the models as an interpretation of the HAADF images because (i) the terrace models have a different number of atoms, (ii) it is unlikely to calculate the exact total terrace area, and (iii) there are numerous candidates for each type of terrace. In the optimized model of small terrace (Fig. 16(a)), one can notice a bonding of Ti (Si) atoms in Ti_3SiC_2 to Si atoms in SiC at the hollow site.

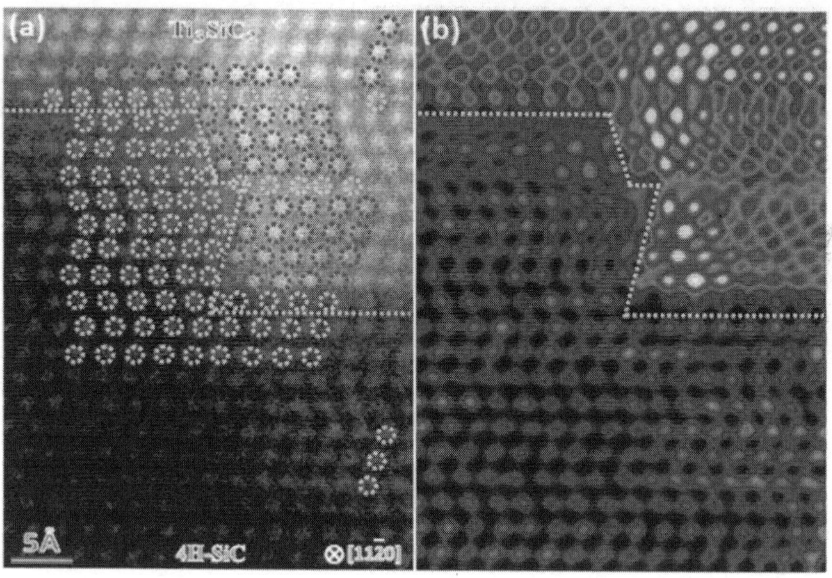

Figure 15. A typical HAADF image of a large terrace observed from the [11-20] direction. The terrace is indicated by a zigzag line. (b) The same image as in (a) but has been filtered to reduce noise [20].

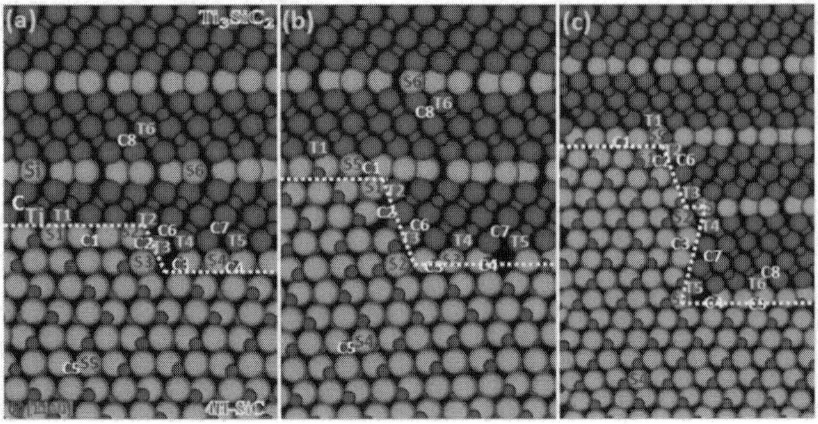

Figure 16. The relaxed atomic models of the three representative terraces with different dimensions: (a) small, (b) intermediate, and (c) large terrace. The representative atoms surrounding terrace are labeled for the atom-projected density of states analysis [20].

Figure 17 shows PDOS of several representative atoms at the small terrace (labeled in Fig. 17(a)), where a remarkable difference is seen between the atoms near and away from terrace. A key feature of this figure is that strong hybridization takes place between the Ti *d* (T1~T6) and Si *p* (S1~S6) levels below E_F, which continues well into the SiC region, inducing a pronounced gap state at E_F for the C atoms near the terrace (C1~C4, C6). However, such a gap state at E_F is vanished completely for the C deeper into the SiC (C5 in Fig. 17). Apart from the C, the Si atoms in SiC near terrace (S1~S3) also exhibit a weak yet visible peak at E_F that are totally absent in the bulk (S5 in Fig. 17). The presence of gap states can be attributed to the overlap of the Ti *d* with Si *p* levels. A notable hybridization is seen as well between the Si *sp* (S1~S4) and C *sp* (C1~C4, C6) levels, indicating the formation of covalency near terrace. Apart from the SiC, all of the atoms in Ti_3SiC_2 around terrace (C6, C7, S4, T1~T5) also exhibit a notable peak at E_F owing to the great degree of overlap between the Ti *d* and C (Si) *p* levels. Moreover, overall feature of PDOS for the Ti atoms around terrace (T1~T4) differs from that in the Ti_3SiC_2 bulk (T6), indicating that the terrace can have an impact on electronic states of Ti as well.

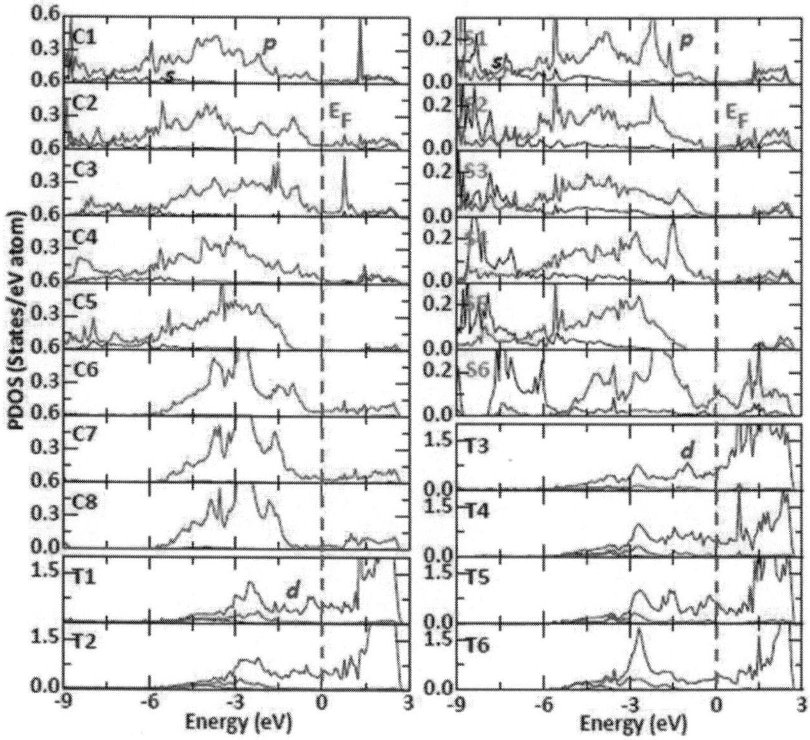

Figure 17. PDOS (states/eV atom) of several selected atoms near or far from the small terrace. All of atoms in the slab are fully relaxed. Refer to Fig. 16(a) for the sites of the selected atoms. The E_F is set to zero and indicated by a vertical dashed line [20].

To identify the bonding nature directly, we further show the contour plots of charge density and density difference for the optimized small terrace viewed along the (11-20) plane (Fig. 18). From Fig. 18(a), a remarkable difference is observed in charge distribution on C: charge distribution around C in SiC (away from terrace) has humps directed toward neighboring Si, while that around C in Ti_3SiC_2 (away from terrace) is of almost spherical symmetry. However, the charge distribution on some C atoms near the terrace (indicated by arrows in Fig. 18(a)) shows a mixed character with a lobe on one side while a spherical outline on the other, which is reflected from their different PDOS (Fig. 17). Furthermore, the charge distribution on the Si-C bonds closet to the zigzag line (Fig. 18(a)) shares some features with that on the Si-C bonds away from the zigzag line (analog to the bonds in SiC bulk): (i) the majority of charges are distributed on all the C atoms, and (ii) there are visible distortions in the charge distribution on the C atoms directed toward their adjacent Si atoms. A certain level of covalency is seen on the atomic bonding along the zigzag line (which defines the terrace), which is due to the hybridization of Ti *d* with Si (C) *p* levels (Fig. 17). These imply a mixed covalent-ionic bonding for the small terrace. It is obvious from the density-difference plot (Fig. 18(b)) that the ionic nature arises from the charge transfer of Ti (Si) to C. In the Ti_3SiC_2 region away from the zigzag line, the Ti-C bonds are found to have more covalent element than the Ti-Si bonds, as more amount of charges are accumulated in between Ti and C, which indicates that the Ti-C is a stronger chemical bond than the Ti-Si.

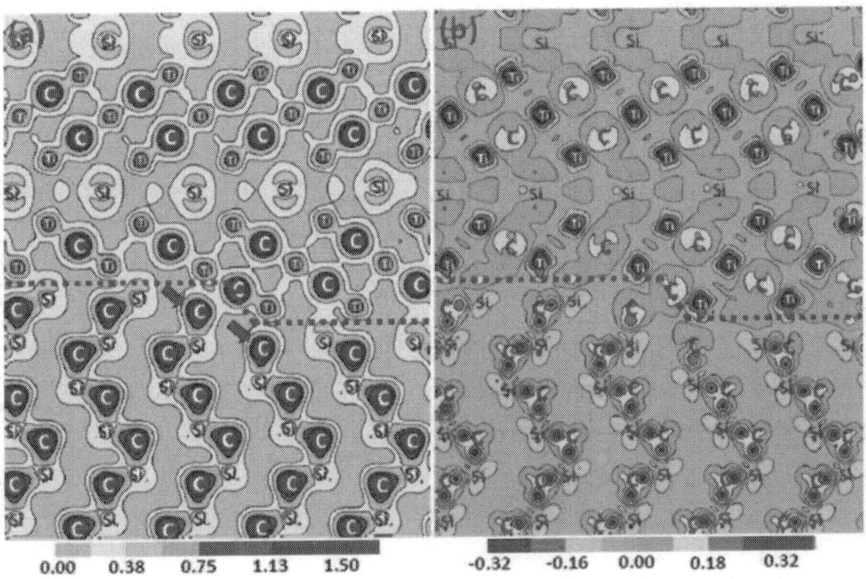

Figure 18. Contour plots of (a) charge density and (b) charge-density difference for the small terrace viewed along the (11-20) plane. Difference in charge density shows redistribution of charge near terrace relative to its isolated system. The position of terrace is indicated by a zigzag dashed line and the atoms intersecting the contour plane are labeled [20].

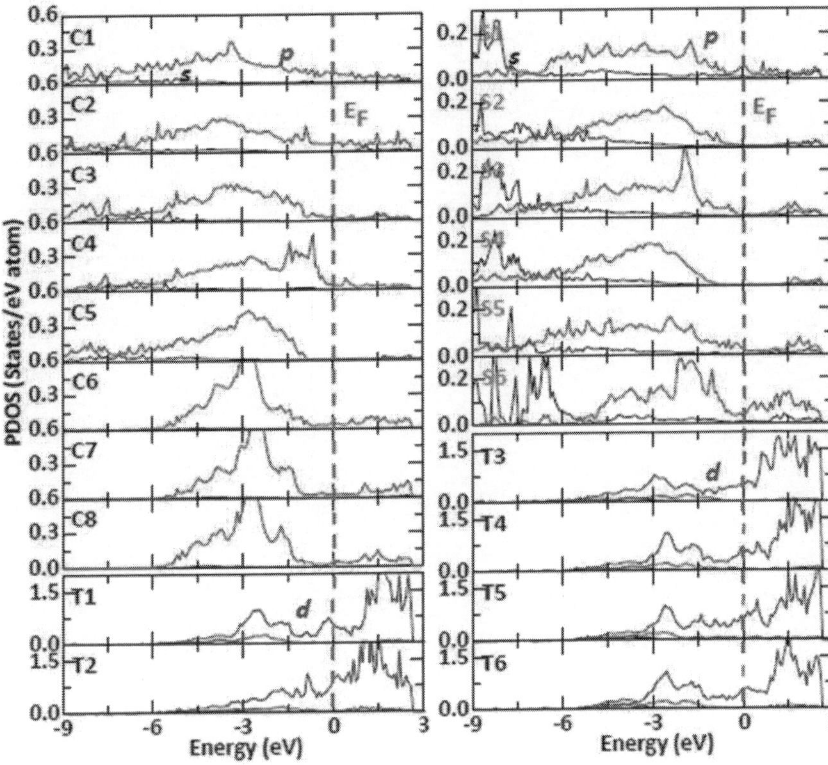

Figure 19. PDOS (states/eV atom) of several selected atoms at or far from the intermediate terrace. Refer toFig. 16(b) for the sites of the selected atoms [20].

The fully relaxed structure of the intermediate terrace is shown in Fig. 16(b), where one can see a Si-Si bonding at the hollow site (on two sides of the zigzag line). Figure 19 shows PDOS plot of several representative atoms on or near terrace (labeled in Fig. 16(b)), where one can note that (i) gap states atE_F appear for the C (C2~C4) and Si (S1,S2) atoms in SiC near terrace, (ii) a hybridization is observed between the Ti *d* (Si *p*) and O *p* levels just below E_F, an indication of the formation of covalency, and (iii) the impact of terrace on the electronic states of both SiC and Ti_3SiC_2 is confined, as the PDOS returns to its bulk value for the atoms away from terrace (*e.g.*, C5, C8, S4, S6, T6). Interestingly, overall feature of the PDOS for either the C (C1~C4) or Si (S1~S3) atoms around terrace differs from one another, suggesting a strong effect of terrace on electronic states of both SiC and Ti_3SiC_2.

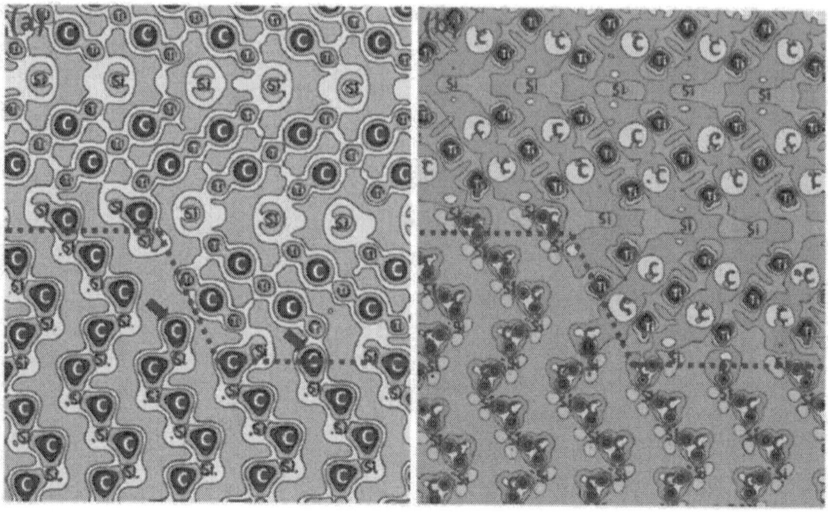

Figure 20. Contour plots of (a) charge density and (b) charge-density difference for the intermediate terrace viewed along the (11-20) plane. The scale is the same as in Fig. 18. The location of terrace is indicated by a zigzag line and the atoms intersecting the contour plane are labeled [20].

Figure 20 shows contour plots of charge density and density difference for the relaxed intermediate terrace intersected along the same plane as in Fig. 18. Like what was seen in the small terrace, the majority of charges remain concentrated on C in two different ways: the charge distribution on the C in Ti_3SiC_2 (*e.g.*, C8) is of spherical symmetry, while that on the C in SiC (*e.g.*, C5) has notable lobes pointed toward their adjacent Si. However, some C atoms near terrace (indicated by arrows in Fig. 12(a)) have lobe and sphere outline simultaneously. Moreover, charges are distributed along the bonds near terrace (on two sides of zigzag line), which together with the charge localization on C atoms infers that the intermediate terrace has a mixed covalent and ionic bonding as well. Evidently, the charge gain on C is at expense of charge loss on their neighboring cations (Fig. 20(b)).

Figure 16(c) illustrates optimized atomic geometry of the large terrace, where a Si-Si bonding is revealed. Figure 21 shows the PDOS of the large terrace, where one can notice that electronic structure is influenced remarkably by terrace. The key point is that there emerge notable peaks at E_F for both the C (C1~C5) and Si (S1~S3, S5, S6) surrounding terrace.

However, such gap states are screened rapidly, since the atoms in SiC away from terrace (S4) show no peak at E_F at all, which implies that the effect of terrace is confined to within a couple of layers. Moreover, a substantial hybridization is seen between the Ti *d* and Si (C) *p* levels below E_F, which indicates the formation of covalent bonding. The covalency is also reflected in the charge-density plot (Fig. 22(a)) showing lobes for the C in SiC and charge accumulation along the bonds in Ti_3SiC_2. These, along with the charge distribution on C at expense of Ti and Si (Fig. 22(b)), indicate that bonding near terrace is again primarily ionic, yet maintains a certain degree of covalency. Finally, both the electron density and density difference for some atoms near terrace (indicated by arrows) deviate severely from those for their bulk counterparts, implying an essential electronic role of the Ti_3SiC_2 in the semiconductor.

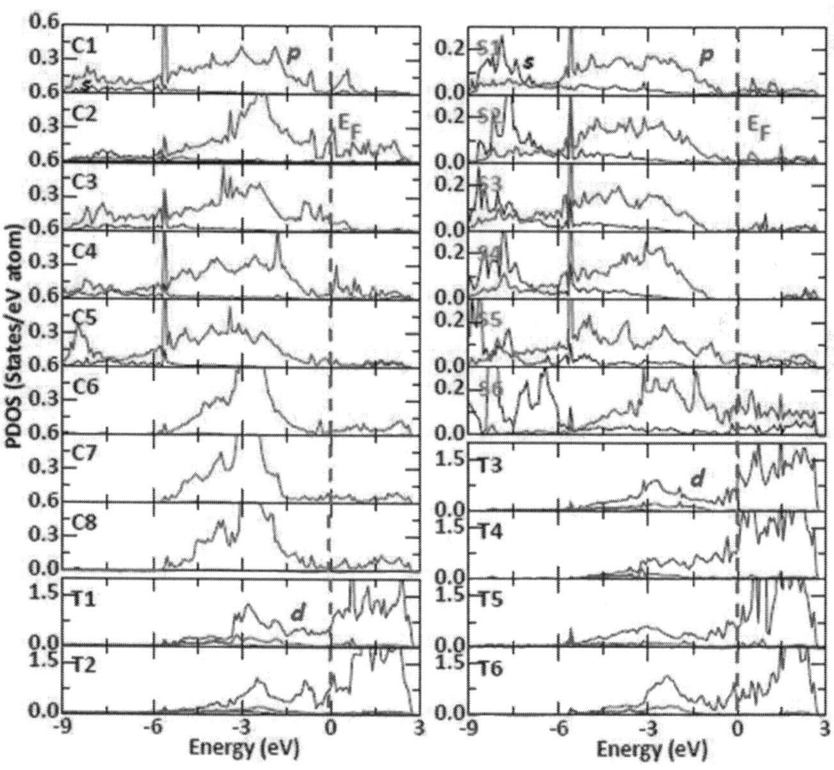

Figure 21. PDOS (states/eV atom) of several selected atoms at or far from the large terrace. Refer to Fig. 16(c) for the sites of the selected atoms [20].

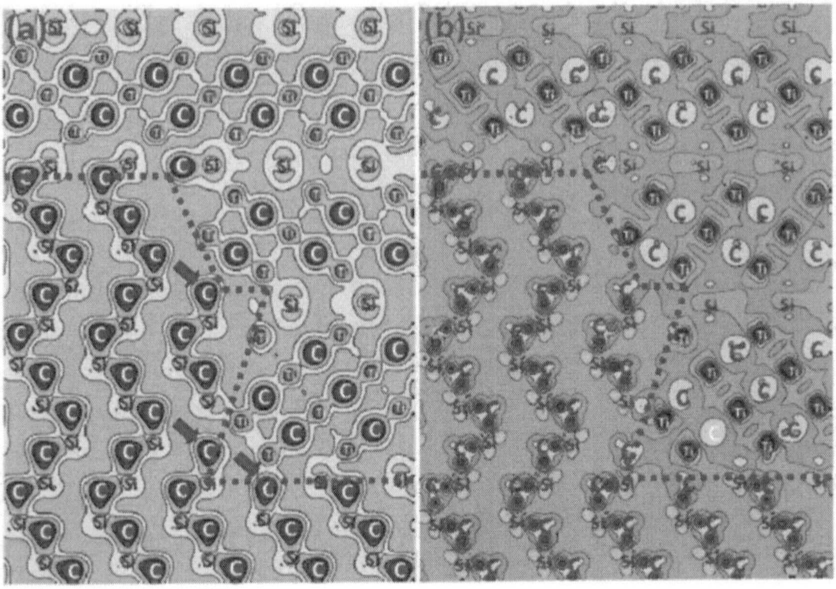

Figure 22. Contour plots of (a) charge density and (b) charge-density difference for the large terrace viewed along the (11-20) plane. The scale is the same as in Fig. 18. The site of terrace is indicated by a zigzag line and the atoms intersecting the contour plane are labeled [20].

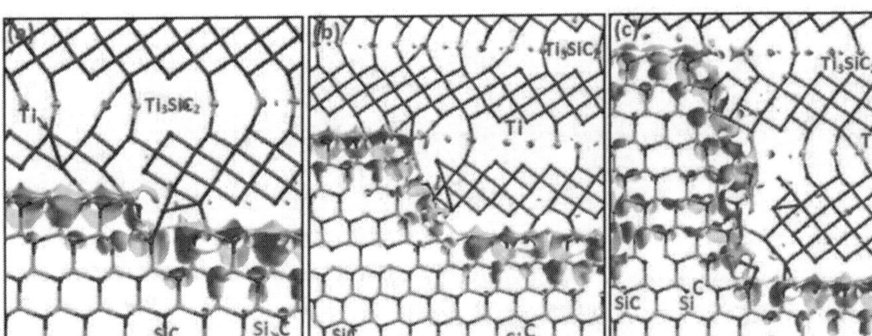

Figure 23. Isosurface plots for (a) small, (b) intermediate, and (c) large terrace along the (11-20) plane in an energy window (E_F–0.5 eV, E_F). The charges on Ti_3SiC_2 are omitted for clarity [20].

4.3. Electron Distribution Near Fermi Energy

Although the PDOS and charge analyses can reveal valuable information on the bonding nature near terrace, they provide limited insights into matters regarding electron distribution around E_F, which is strongly related to the electronic conduction over terrace. Figure 23 shows electron-density isosurface near E_F along the (11-20) plane for the three terraces. The electrons are accumulated substantially around all the terraces in a spatially connected fashion, which are extended as far as several atomic layers into SiC, regardless of dimension of the terraces. Such a broad electron distribution provides a likely electron conduction channel to allow current transport across a few layers of the semiconductor, which indicates that terraces could also be one of the origins underlying the observed ohmic nature in the metal/SiC contact system. One can also note that the three terraces are comparable in the amount of accumulated charge at E_F, inferring that the dimension of terrace plays an insignificant role in affecting electron transport across the contact. It is worthy of mentioning that the electron density at E_F is extremely high in Ti_3SiC_2 (sea of electrons, not shown for clarity), yet turns almost nil for the SiC atoms away from terrace, which can be understood on the fact that the Ti_3SiC_2 is intrinsically metallic, whereas the SiC is semiconducting.

5. DISCUSSION AND CONCLUSIONS

The current understanding of formation origin of Ohmic contact, which is based mainly on experimental studies of property measurement and structure characterization, can be summarized in three main points [27-30]: (1) the deposited Al (80 at%) might diffuse in part into the SiC and dope heavily the semiconductor because Al is well-known to act as a *p*-type dopant for SiC, (2) the high density of pits, spikes, or dislocations may be generated underneath the contacts after annealing so that current can transport primarily through these defects due to the possible enhancement of electric field at these features and semiconductor doping at these locations, (3) formation of intermediate semiconductor layer between the deposited metals and semiconductor, which consists of silicides or carbides, could divide the high barrier height into lower ones, thus reducing the effective barrier height.

The findings presented first demonstrate that no Al is clearly segregated around the interfacial region, in particular at the top few layers of SiC, which rules out the possibility of additional Al doping. Though a small amount of residual Al is found to be present, mostly in a form of Al_4C_3 compound, it may locate on the surface of annealed contacts rather than in the layer directly contacted to the SiC, thus playing a negligible role in Ohmic contact formation. The majority of deposited Al is evaporated during annealing because of its low melting point and high equilibrium vapor pressure. The dominant role played by Al in the TiAl system is to assist the formation of liquid alloy so as to facilitate chemical reaction. Furthermore, careful characterization of the interfacial region reveals that the substrate and the generated compound are epitaxially oriented and well matched at interface with no clear evidence of high density of defects. This suggests that the morphology might not be the key to understanding the contact formation. In support of this speculation, it has been observed previously that Ti Ohmic contacts can be possibly generated without any pitting and that pit-free Ohmic contacts can be fabricated.

One remaining theory is the alloy-assisted Ohmic contact formation. This alloy is determined to be ternary Ti_3SiC_2, which has also been corroborated by other expriments. Since the bulk Ti_3SiC_2 has already been found to be of metallic nature both in experiment and theory, the contact between $Ti_3SiC_2$and its covered metals should show Ohmic character and thus the SiC/Ti_3SiC_2 interface should play a significant role in Ohmic contact formation. This idea is supported by the fact that the determined interface has a lowered SBH due to the large dipole shift at interface induced by the partial ionicity and the considerable charge transfer. In addition, the interfacial states, as indicated by the electron distribution at $E\ F$, are also viewed as a contributing factor in reducing the SBH. These states might be further enhanced by the possible presence of point defects at interface, although these structural defects have not been detected by the TEM study.

Interestingly, our calculations predict that an atomic layer of carbon emerges as the first monolayer of Ohmic contacts, which eventually affects interface electronic structure. Such trapped carbon was previously studied in both other interfacial systems theoretically by DFT and the TiNi Ohmic contacts on 4H-SiC experimentally by Auger electron spectroscopy. It was proposed that the carbon could be segregated to the interfacial area, strengthening interface substantially and reducing the Schottky barrier dramatically. Further, it was reported that the Ohmic contact can be realized by depositing carbon films only onto the SiC substrate, indicative of the determinative role of carbon in the Ohmic contact formation. The important role played by carbon in our study can be traced to the two interfacial Si layers, which provide possible sites for carbon segregation due to the strong Si-C interaction. However, direct imaging of the trapped carbon is still difficult in present study and further characterization requires the high-voltage EM and/or other advanced microscopic techniques.

We then demontrate that atomic-scale Ti_3SiC_2-like bilayer can be embeded in the SiC interior, forming an atomically ordered multilayer that exhibits an unexpected electronic state with the point Fermi surface, in stark absence in repestive bulk constituents. The valence charge is found to be confined largely within the bilayer in a spatially connected way, which serves as a possible conducting channel to enhance the current flow over the semiconductor. Such a heterostructure with unusual properties is mechanically robust, rendering its patterning for technological applications likely. Finally, the atomic structures of terraces at the contacts in SiC devices are investigated and bridged to their electronic properties at an atomic scale. Experimentally, newly formed carbide Ti_3SiC_2 is demonstrated to bond directly to silicon carbide in the terrace region in an epitaxial and atomically ordered fashion, regardless of dimension of terraces. Further first-principles calculations reveal gap states in the semiconductor layers and a substantial charge accumulation around terraces in a connected and broadly distributed manner. The presence of gap states at Fermi energy and the likelihood to serve as electron conduction channels to allow current flow over the semiconductor identify the terraces as one of the origins underlying the ohmic contact in silicon carbide electronics. Such a combined experimental and theoretical investigation provides insight into the complex atomic and electronic structures of buried terraces, which should be applicable to addressing contact issues of interest in other electronic devices.

To summarize, we have determined in this chapter atomic-scale structures of Ohmic contacts on SiC and related them to their electronic structures and electric properties, aimed at understanding the formation mechanism of Ohmic contact in TiAl-based system. The combined HAADF-DFT study [31] represents an important advance in relating structures to device properties at an atomic scale and is not limited to the contacts in SiC electronics. Our results show that the main product generated by chemical reaction can be epitaxial and have atomic bonds to the substrate. The contact interface, which could trap an atomic layer of carbon, enables lowered Schottky barrier due to the large interfacial dipole shift associated with the considerable charge transfer. The atomic-scale Ti_3SiC_2-like bilayer is embedded well in SiC bulk interior in an epitaxial, coherent and atomically abrupt manner, which exhibits an unexpected state with a point Fermi surface.

Moreover, the formed Ti_3SiC_2 can even be epitaxial and atomically ordered on SiC substrate near terrace, which inudces pronounced gap states at E_F in the semiconductor layers. Charges are accumulated heavily surrounding terrace in a spatially connected fashion, irrespective of dimension of the terraces, which suggests the possiblity of terraces as likely electron conduction channels to allow current transport across the semiconductor. The inducing of gap states and the capability to enable current flow over the semiconductor identify the terraces as one of the origins underlying the Ohmic nature in the metal/SiC contact system as well. These findings are relevant for technological improvement of contacts in SiC devices, and this chapter presents an important step towards addressing the current contact issues in wide-band-gap electronics.

ACKNOWLEDGEMENTS

The author acknowledges M. Saito, S. Tsukimoto, at the WPI Research Center, Advanced Institute for Materials Research, Tohoku University and Y. Ikuhara at the The University of Tokyo for their collaborations. The author thanks S. Watanabe at The University of Tokyo for allowing our use of computational resources. The present study was supported in part by a Grant-in-Aid for Scientific Research on Priority Area, "Atomic Scale Modification (474)" from the Ministry of Education, Culture, Sports, Science, and Technology of Japan. Z. W acknowledges financial supports from the Grant-in-Aid for Young Scientists (A) (Grant No. 24686069) and the Challenging Exploratory Research (Grant No. 24656376). The calculations were carried out on a parallel SR11000 supercomputer at the Institute for Solid State Physics, Univ. of Tokyo.

REFERENCES

1. I. Perez-Wurfl, R. Krutsinger, J. T. Torvik, B. Van Zeghbroeck, 2003 4H-SiC Bipolar Junction Transistor with High Current and Power Density Solid State Electronics 47 229 231

2. S. K. Lee, C. M. Zetterling, E. Danielsson, M. Östling, J. P. Palmquist, H. Högberg, U. Jansson, 2000 Electrical Characterization of TiC Ohmic

Contacts to Alumimum Ion Implanted 4H-Silicon Carbide Applied Physics Letters 77 1478 1480

3. B. Pécz, 2001 Contact Formation in SiC Devices Applied Surface Science 184 287 294

4. J. Crofton, P. A. Barnes, J. R. Williams, 1993 Contact Resistance Measurements on p-type 6H-SiC Applied Physics Letters 62 384 386

5. S. E. Mohney, B. A. Hull, J. Y. Lin, J. Crofton, 2002 Morphological Study of the Al-Ti Ohmic Contact to p-type SiC Solid State Electronics 46 689 693

6. O. Nakatsuka, T. Takei, Y. Koide, M. Murakami, 2002 Low Resistance TiAl Ohmic Contacts with Multi-Layered Structure for p-Type 4H-SiC Materials Transactions 43 1684 1688

7. B. J. Johnson, M. A. Capano, 2004 Mechanism of Ohmic Behavior of Al/Ti Contacts to p-type 4H-SiC After Annealing Journal of Applied physics 95 5616 5620

8. S. C. Chang, S. J. Wang, K. M. Uang, B. W. Liou, 2005 Investigation of Au/Ti/Al Ohmic Contact to N-type 4H-SiC Solid State Electronics 49 1937 1941

9. T. Ohyanagi, Y. Onose, A. Watanabe, 2008 Ti/Ni Bilayer Ohmic Contact on 4H-SiC Journal of Vacuum Science and Technology B 26 1395 1362

10. S. Tsukimoto, K. Nitta, T. Sakai, M. Moriyama, M. Murakami, 2004 Correlation Between the Electrical Properties and the Interfacial Microstructurs of TiAl-Based Ohmic Contacts to p-type 4H-SiC Journal of Electronic Materials 33 460 466

11. Z. C. Wang, S. Tsukimoto, M. Saito, Y. Ikuhara, 2009 SiC/Ti3SiC2 Interface: Atomic Structure, Energetics, and Bonding Physical Review B 79, 045318 1 10

12. P. D. Nellist, M. F. Chisholm, N. Dellby, O. L. Krivanek, M. F. Murfitt, Z. S. Szilagyi, A. R. Lupini, A. Borisevich, W. H. Sides Jr, S. J. Pennycook, 2004 Direct Sub-Angstronm Imaging of a Crystal Lattice Science 305 1741

13. Z. C. Wang, M. Saito, K. P. Mc Kenna, L. Gu, S. Tsukimoto, A. L. Shluger, Y. Ikuhara, 2011 Atom-Resolved Imaging of Ordered Defect Superstructures at Individual Grain Boundaries Nature 479 380 383

14. Z. C. Wang, S. Tsukimoto, M. Saito, Y. Ikuhara, 2009 Atomic and Electronic Structure of the YBa2Cu3O7/SrTiO3 Interface from First Principles Journal of Applied Physics 106 093714-1 8

15. Z. C. Wang, S. Tsukimoto, M. Saito, Y. Ikuhara, 2009 Individual Charge-Trapping Dislocations in an Ionic Insulator Applied Physics Letters 95 184101-1 3

16. Z. C. Wang, W. Zeng, L. Gu, M. Saito, S. Tsukimoto, Y. Ikuhara, 2010 Atomic-Scale Structure and Electronic Property of the LaAlO3/TiO2 Interface Journal of Applied Physics 108 113701-1 9

17. Z. C. Wang, M. Saito, S. Tsukimoto, Y. Ikuhara, 2009 Interface Atomic-Scale Structure and its Impact on Quantum Electron Transport. Advanced Materials 21 4966 4969

18. Z. C. Wang, S. Tsukimoto, M. Saito, K. Ito, M. Murakami, Y. Ikuhara, 2009 Ohmic Contacts on Silicon Carbide: The First Monolayer and its Electronic Effect Physical Review B 80 245303 1 12

19. Z. C. Wang, S. Tsukimoto, R. Sun, M. Saito, Y. Ikuhara, 2011 Atomic-Scale Ti3SiC2 Bilayers Embedded in SiC: Formation of Point Fermi Surface Applied Physics Letters 98 104101-1 3

20. Z. C. Wang, M. Saito, S. Tsukimoto, Y. Ikuhara, 2012 Terraces at Ohmic Contact in SiC Electronics: Structure and Electronic States Journal of Applied Physics 111 113717-1 18

21. G. Kresse, J. Hafner, 1993 Ab initio Molecular Dynamics for Liquid Metals Physical Review B 47 558 561

22. X. G. Wang, J. R. Smith, A. Evans, 2002 Fundamental Inflenence of C on Adhesion of the Al2O3/Al Interface Physical Review Letters 89 286102-1 4

23. Z. C. Wang, M. Saito, S. Tsukimoto, Y. Ikuhara, 2011 Heterointerfaces: Atomic Structures, Electronic States, and Related Properties Journal of the Ceramics Society of Japan 119 783 793

24. S. J. Pennycook, L. A. Boatner, 1988 Chemically Sensitive Structure-Imaging with a Scanning Transmission Electron Microscope Nature 336 565 567

25. W. Y. Ching, Y. N. Xu, P. Rulis, L. Z. Ouyang, 2006 The Electronic Structure and Spectroscopic Properties of 3C, 2H, 4H, 6H, 15R and 21R Polymorphs of SiC Materials Science and Engineering A 422 147 156

26. Z. C. Wang, T. Kadohira, T. Tada, S. Watanabe, 2007 Nonequilibrium Quantum Transport Properties of a Silver Atomic Switch Nano Letters 7 2688 2692

27. M. Gao, S. Tsukimoto, S. H. Goss, S. P. Tumakha, T. Onishi, M. Murakami, L. J. Brillson, 2007 Role of Interface Layers and Localized States in TiAl-Based Ohmic Contacts to p-Type 4H-SiC Journal of Electronic Materials 36 277 284

28. S. Tanimoto, N. Kiritani, M. Hoshi, H. Okushi, 2002 Ohmic Contact Structure and Fabrication Process Applicable to Practical SiC Devices Materials Science Forum 389-393 879 884

29. H. Morkoc, S. Strite, G. B. Gao, M. E. Lin, B. Sverdlov, M. Burns, 1994 Large-band-gap SiC, III-V Nitride, and II-VI ZnSe-based Semiconductor Device Technologies Journal of Applied physics 76 1363 1398

30. W. Lu, W. C. Mitchel, CA Thornton, W. E. , G. R. Landis, S. R. Smith, 2003 Ohmic Contact Behavior of Carbon Films on SiC Journal of the Electrochemical Society 153 G177 G182

31. Z. C. Wang, M. Okude, M. Saito, S. Tsukimoto, A. Ohtomo, M. Tsukada, M. Kawasaki, Y. Ikuhara, 2010 Dimensionality-Driven Insulator-Metal Transition in A-site Excess Non-stoichiometric Perovskites Nature Communications 1 106

7

PECVD AMORPHOUS SILICON CARBIDE (A-SIC) LAYERS FOR MEMS APPLICATIONS

Ciprian Iliescu[1] and Daniel P. Poenar[2]

[1] Institute of Bioengineering and Nanotechnology, Singapore,

[2] Microelectronics Centre, School of Electrical & Electronics Engineering, Nanyang Technological University, Singapore,

1. INTRODUCTION

Silicon carbide (SiC) became an important material whose popularity has been constantly increasing in the last period due to its excellent mechanical, electrical, optical and chemical properties, which recommend it in difficult and demanding applications.

There are two main fields of applications of SiC. The first one is related to nano electronic [1] or even integrated circuits [2] which are using SiC (in monocrystalline or –sometimes- polycrystalline form) as basic structural material for high frequency [3], high power [4], high voltages [5], and/or high temperature devices [6] or combinations thereof [7]. In most of these applications, SiC act as are placement material for silicon which cannot be used under such extreme conditions.

The second area of applications is related to sensors [8] and actuators, i.e. structures, devices, and/or Microsystems realized (or at least embedding some elements of) micro- and nano electro mechanical systems (MEMS & NEMS). In this area, the usage of SiC (but mainly in polycrystalline and amorphous form) is mainly due to its compatibility and easy integrability with Si and Si-based microfabrication technology. In this direction, a lot of miniaturized devices, such as chemical sensors

[9], UV detectors [10], MEMS devices [11-13], and even NEMS [14], are using SiC thin films or substrates (6H-SiC or 4H-SiC).

Polycrystalline-SiC (3C-SiC) thin films can be heteroepitaxially grown on Si substrates [14] due to the deposition temperature well below the Si melting point. [15] However, most of MEMS applications require thin films deposition methods at lower temperatures. This is necessary in order to ensure an overall low thermal budget for the entire fabrication of the SiC-based device(s), an essential prerequisite for postprocessing of MEMS structures on top of Si CMOS circuits, which ensures implementation of 'smart sensors'. Plasma enhanced chemical vapor deposition (PECVD) of SiC in an amorphouos state (α-SiC) can be such a solution. Early studies have been done on the structural, optical and electronic properties of this material [16, 17]. More specifically, one of the key challenges for PECVD of SiC for MEMS and NEMS applications is achieving alow residual stress together, if possible, with a high deposition rate and good uniformity [16], [18-20].

The main advantages of PECVD α-SiC deposition can be summarized as follows:

- Low temperature deposition, usually between 200-400^{0}C (depending on the specific s of the machine and recipe employed for the deposition, as well as on the details of the device's fabrication process).
- As a direct consequence of the advantage mentioned previously, α-SiC is a highly suitable structural layer for surface micro machining applications using polyimide [21], amorphous silicon [22] or SiO_2 [23] as sacrificial layer materials.
- Stress control in a wide range (e.g. between −1200MPa and 400MPa [24] by tuning of deposition parameters [25], doping [26] or annealing [27].
- The ability of the fabricated device to operate at relatively high temperatures.
- Large mechanical strength of the deposited α-SiC layer.
- Wide bandgap for the deposited α-SiC layer, making it an almost ideal optoelectronic material, transparent for all visible wave lengths above 0.5μm and thus highly suitable for guiding light in the visible and infrared optical spectrum [28].

- A refractive index greater than 2.5 (significantly larger than that of SiO_2 and even that that of Si_3N_4) also make α-SiC an excellent candidate for optical waveguides [29].

- Capability to deposit conformal layers [30].

- The deposited α-SiC has a coefficient of thermal expansion (CTE) relatively similar with that of Si. This means that the risks of both developing very high internal stress within the film and, therefore, of subsequent delamination, are minimized [31] when the devices are operated even at high temperatures.

- The deposited α-SiC is highly suitable for applications requiring a high level of corrosion resistance and moderate operating temperatures (below 300^{0}C) [32].

- This chapter will focus on the PECVD deposition of α-SiC layers for MEMS/NEMS applications. The chapter is organized in three major parts:

- A detailed description of the typical PECVD reactor and of the important aspects necessary for a competitive α-SiC deposition process.

- The influence of main parameters on the deposition process and how one could achieve a low stress α-SiC film with anexcellent uniformity and a good deposition rate

- Post-deposition processing of the PECVD α-SiC layerfor various devices and applications.

2. PECVD REACTORS

The deposition of α-SiC layers in a Plasma Enhanced Chemical Vapor Deposition (PECVD) reactor is facilitated by the plasma generated between two electrodes (radio frequency-RF- or DC discharge) in the presence of reacting gases, the substrate being connected at one of these electrodes. There is a very large diversity of types of PECVD reactors on the market, either for industrial applications or specially designed for dedicated R&D. Usually the R&D equipments are more complex but also may allow much more of flexibility and degrees of freedom in controlling the deposition, thus enabling to obtain different layers with distinct properties using the same reactor.

The key elements in the selection of a PECVD reactor can be summarized as follows:

Deposition chamber. The operating temperature for most of the PECVD reactors is between 200-400°C. In order to achieve a uniform deposition special attention has to be paid to the inlet of the reactive gases, which can be of three types: several inlets around the bottom chuck (electrode), or one inlet through the top electrode, or multiple inlets (shower) from the top electrode. The last solution seems to ensure a better distribution of reactive gases between the working electrodes with positive effect on the film's uniformity. Meanwhile, preheating the gases (using a heated gas distribution system) before their actual introduction into the deposition chamber may also improve the deposition uniformity. Heating the deposition chamber itself (usuallyat 50-100°C) generates gradients of temperature that avoid particle deposition on the substrate during processing. Of course, adding all these elements into a standard system would finally be reflected in a higher cost of the tool.

Loadlock system. Two types of reactors can be distinguished, depending on whether a loading system is present or not: open systems (without lock load, i.e. relatively cheap reactors used only in research labs) and closed systems. The presence of a vacuumed loading system is also a critical element for good PECVD deposition, for a stable and repetitive process. There are two main aspects related to the presence of the load lock: one is related to safety while the other one is related tothe quality of the deposited layer. For the safety aspect, the presence of the loading system avoids the contact of the operator with the by-products resulted during processing, some of which are carcinogenic. As for the process quality, the fact that the chamber is kept permanently under vacuum results in excellent film quality with outstanding reproducibility.

Reconfiguration of the chamber. Cleaning of the chamber is also an important element in achieving good-quality layers. Most of the PECVD reactors also allow a „plasma cleaning "process, which is applied once a certain thickness of the deposited layer is achieved (usually 5-10μm). This cleaning process is designed to remove the products deposited on the chamber's walls or on electrodes, and it is performed mainly using CF_4/O_2 or C_4F_8/O_2 as reactive gasses. The process is followed by a short pre-deposition of the material desired to be deposited in the reactor. Mechanical cleaning must be also performed periodically.

Gas precursors. The PECVD systems frequently used in R&D are equipped with a large number of inlets for the reactive gases. In most of the cases, the equipment is used for multiple depositions such as SiO_2 (doped and undoped), Si_3N_4, α-Si or even TEOS (using a special Liquid Delivery System –LDS). In our case, for the deposition of α-SiC layers, silane (SiH_4) and methane (CH_4) are the most often used gas precursors, although other precursors, e.g. methyltrichlorosilane (MTCS) [33] or SiH_4/acetylene (C_2H_2) [34], were also studied.

3. INFLUENCE OF THE DEPOSITION PARAMETERS

3.1. Materials and Methods

This section describes the influence of the main deposition parameters on the film properties, being a practical guide of parameter selection for a desired characteristic of the deposited α-SiC thin film. The experiments were performed on 4 inch diameter, *p* type, 500μm-thick silicon wafers with a (100) crystallographic orientation. The wafers were initially cleaned in piranha (H_2SO_4:H_2O_2 in the ratio of 2:1) at 120°C for 20 minutes and rinsed in DI water. The native silicon oxide layer was then removed by immersing the wafers in a classical BOE solution for 30 seconds. Stress measurement was performed using a KLA Tencor FLX-2320 system while the thickness and thickness uniformity of the thin filmswere measured with a refractometer (Filmetrics F50, USA). The deposition of the tested α-SiC layers was performed using a STS Multiplex Pro-CVD PECVD system described in detail elsewhere [35, 36]. This system enables two RF deposition modes: a low frequency (LF) mode at 380 kHz with a tuning power between 0 to 1 kW, and/or a high frequency (HF) mode at 13.56 MHz with a selected power in the range between 0 to 600 W. The depositions of the α-SiC layers were performed using pure SiH_4(pure) and CH_4 as precursors, with Ar as an overall dilution gas.

3.2. Pressure

The α-SiC deposition‘s uniformity is strongly dependent on the pressure in the reactor chamber (Fig. 1a).

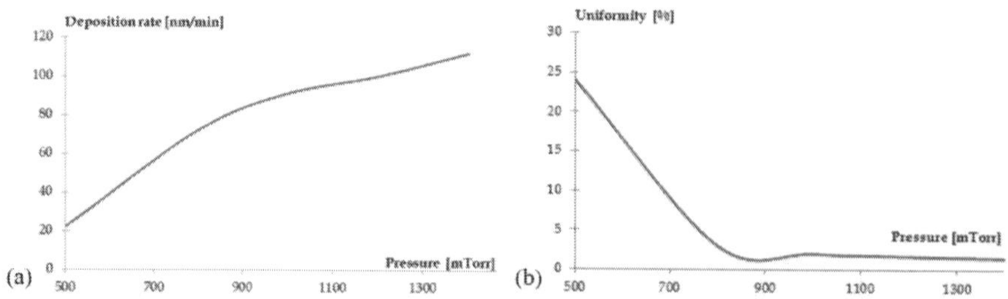

Figure 1. Variation of: (a)uniformity, and (b)deposition rate, with the pressure for α-SiC PECVD deposition in a STS Multiplex Pro-CVD PECVD system.

Below 800mTorr, large non-uniformity values were observed, but the deposition uniformity linearly improved as the pressure was varied between 500 to 800mTorr, finally settling to a constant valueof the uniformities under 2% for all the pressures in the range 900 to 1400mTorr. The thickness uniformity‘s map presented a "donut" shape, which means that the gas molecules present a high velocity, increasing the deposition rate at the edge of the wafer.

Another important aspect of the pressure is its influence on the deposition rate (Fig. 1b): low pressures reduce the concentration of reactive species thus resulting in a low deposition rate, which increases quasi-linearly with pressure.

3.3. Rf Power

Both the RF power and the power deposition mode are key parameters for tuning the optical and mechanical properties of the deposited PECVD α-SiC layer. A low value of the residual stress is required in MEMS applications where free standing structure is fabricated. Meanwhile, a high deposition rate is desired mainly in industrial applications. Fig. 2 illustrates the variation of the deposition rate and residual stress versus the RF power for the HF mode. The linear dependence of the deposition rate on the HF power noticed for power levels below 300W can be explained by the proportional increase in the dissociation of reactant gases with increasing the power. After a ‘threshold’ value (300W) the dissociation into reactive species is no longer a crucial factor as probably all reactive species are easily and fully dissociated, hence further increasing the power has little effect on the deposition rate.

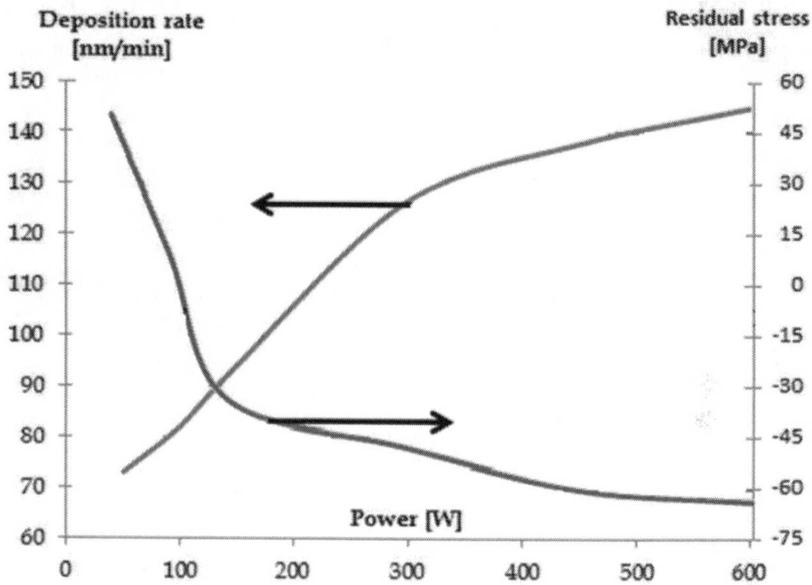

Figure 2. Variation of the deposition rate and residual stress with the HF power in the STS Multiplex Pro-CVD PECVD reactor. The deposition temperature was 300°C, the pressure 1100mTorr, and the gas flow rates of SiH_4, CH_4 and Ar were 45, 300 and 700sccm, respectively.

An interesting characteristic of PECVD α-SiC thin films deposited using the dual mode technique, particularly when compared to other PECVD deposition methods, is the very low value of the internal average stress, which can range between 50MPa and −70MPa. Fig. 2 shows the variation of this residual average with the RF power. Our results indicate that the average stress variation of a α-SiC film deposited in the HF deposition regime is not due to modifications of the film's chemical composition, but rather due to internal structural re-arrangement because the increasingly higher values of HF power quickly provide enough energy for the adsorbed species to migrate and find the more energetically favorable sites for the film growth to take place.

The refractive index was almost constant in the range between 2.5 and 2.6 while the uniformity of the deposition and uniformity of the refractive index was below 1.5%.

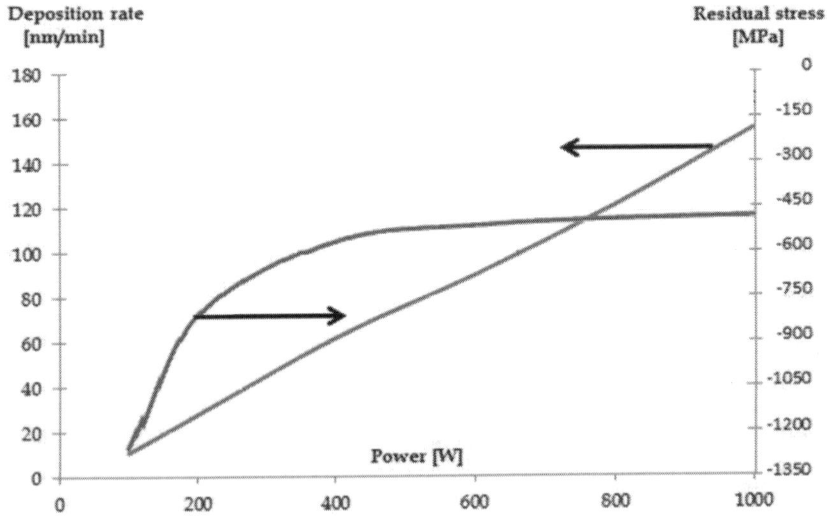

Figure 3. Variation of the deposition rate and residual stress with the LF power in the same PECVD reactor under the same conditions as in Fig. 2.

The variation of the deposition rate and residual stress of PECVDα-SiC films deposited in LF mode are presented in Fig. 3. The differences between HF mode and LF mode are quite evident if one compares the results shown in Fig. 3 with those shown in Fig. 2. The deposition rate depends linearly with the LF power. However, for low values of LF power, the stress is highly compressive (even below −1200MPa) but quickly reduces to about −500MPa and remains almost constant at this value for any LF power above 300W. This variation can be explained by the densification of the layer determined by the increasingly more energetic ion bombardment with increasing the LF power, which characterizes the LF deposition mode. At high frequencies, only the electrons are able to follow the RF field while the ions cannot follow the instantaneous variations of the electric field due to their heavier mass. The cross over frequency at which the ions start following the electric field is between 1 and 5MHz depending upon the mass of the ions.

Consequently, below 1MHz, the ion bombardment is significantly higher, which not only enhances chemical reactions but also densifies the film. However, the refractive index of α-SiC film decreases with the increasing the LF power (from 2.9 to 2.5). It can be concluded that, for a low value of the residual stress in the α-SiC layers, the HF deposition mode is more suitable. Meanwhile, the HF mode assures a better dissociation of the gasses that is reflected in a higher deposition rate.The deposition in LF mode is more suitable when PECVD α-SiC is used as masking layer or in applications for harsh environments (due to the densification of the layer).

3.4. Temperature

A very interesting characteristic of the PECVD α-SiC deposition in the LF mode and which can be useful in many applications is that the deposition rate as well as the residual stress do not present relevant variations as a function of temperature [37]. In contrast, in the HF mode, temperature has an important effect on the stress value, although the deposition rate is not much affected by temperature in HF mode as well [37]. As is depicted in Figure 4, the internal average stress is compressive (around −110MPa) when the α-SiC layers are deposited at 200°C, but it becomes more and more tensile with increasing the deposition temperature,the stress (for temperature between 350 and 400°C). Most remarkable is the fact the stress is very low (around zero) when the deposition temperature is situated between 300 and 350°C.

The refractive index value remained almost constant at 2.6 for all the films deposited in the HF mode which proves that the stress variation is resulted from the film's structural re-arrangement, not from modifications of its chemical composition [37].

3.5. Sih_4/Ch_4 Ratio

The influence of the SiH_4/CH_4 gas flow ratio, illustrated in Fig. 5, shows a linear dependence of the (compressive) residual stress on the SiH_4/CH_4 ratio for depositions in the HF mode. As expected, increasing the SiH_4 content from 1:10 to 2:10 in the gas flow makes the deposited α-SiC film more „Si-rich" and at the same time decreases the magnitude of the compressive stress from about –70MPa to –20MPa, while at the same time the refractive index increases from 2.6 to 2.8.

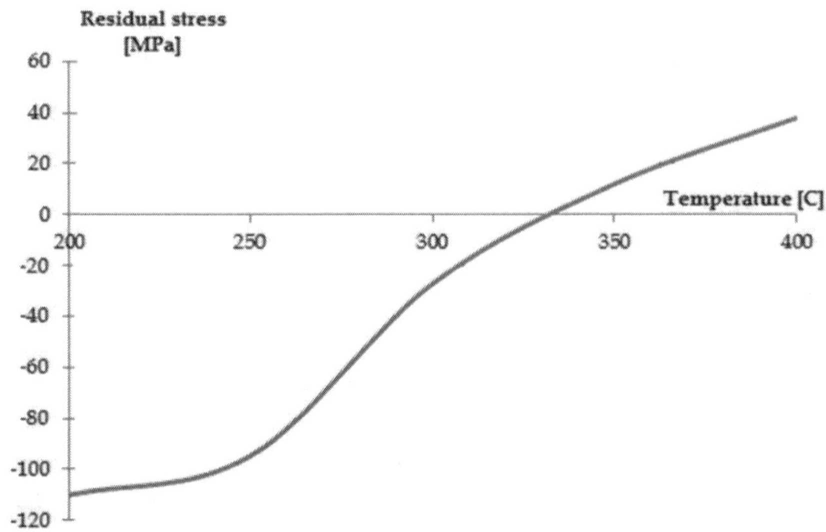

Figure 4. Variation of the average residual stress of PECVD α-SiC films with the deposition temperature. The films were deposited in the HF mode at a power of 150W, at a pressure of 1100mTorr, and with gas flow rates of SiH_4, CH_4 and Ar of 45, 300 and 700sccm, respectively.

4. MEMS APPLICATION OF PECVD A-SIC LAYERS

4.1. Patterning Of Sic Layers

Micro patterning of thin α-SiC film layer deposited in PECVD reactor is similar to the patterning of crystalline or polycrystalline SiC layers. The patterning can be done in Reactive Ion Etching (RIE) or Inductive Coupled Plasma (ICP) deep RIE (DRIE) equipments by fluorine chemistry, i.e. using a fluorine-containing gas (CF_4,SF_6 or even CHF_3) and O_2 and using photoresist as masking layer. It must be mentioned that the selectivity of the etching process to Si (used as substrate), SiO_2, or even to photoresist is not outstanding (ranging from 1.2 up to 0.5) [16]. Metal masking layers have thus been initially been choosen [38, 39]. However, metal masks cause a major problem, namely the micromasking effect: as dry etching progresses, metallic particles are extracted from the mask and re-deposited onto the film/substrate where they can continue to have a protective role, resulting in a very uneven and rough final surface. For instance, aNF_3/O_2 chemistry was tested for etching a SiC layer [40] using to photoresist is mask, achieving etching rates of 0.135μm/min. Similarly, HBr/Cl_2 etching chemistries using SiO_2 etch masks have been developed [41]. This latter chemistry allowed a high selectivity (20:1) of etching SiC with respect to SiO_2, but the price to pay was a very low etching rate (0.02μm/min). In a more recent work, Senesky and Pisano reported the usage of AlN as masking layer for SiC structures in an ICP DRIE reactor using SF_6/O_2 chemistry [42]. The etching process yielded a SiC etch rate of 0.4μm/min, having a high selectivity (SiC/AlN) of 16:1. Moreover, the anisotropy of the process was very good, as features with a sidewall angle of 10° were reported. In another work,Cl_2chemistry was used in an ICP DRIE reactor (by Pandraud et al.) for uniform patterning of a PECVD α-SiC layer for wave guide applications [43].

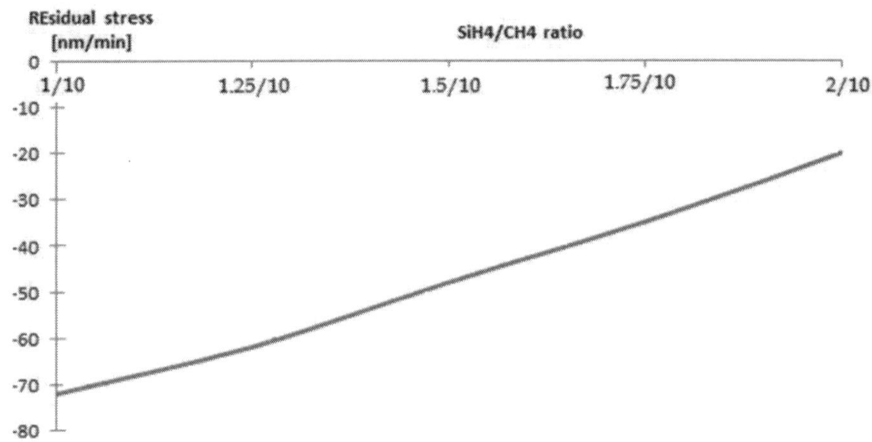

Figure 5. Variation of the average residual stress with the SiH_4/CH_4ratio for PECVD α-SiC films deposited at 300°C. The SiH_4 flow rate was kept constant at 45sccm, while all the other deposition conditions were the same as those indicated in Fig. 4.

To conclude, an optimized process recipe for the deposition of a low stress α-SiC film using a STSP PECVD reactor has the following parameters:

Table 1.

Process parameters/ Characteristic of the deposition process	Value
Pressure	1400mTorr
Power (HF mode)	600W
SiH_4, CH_4 and Ar gas flows	70, 500, and 700sccm
Deposition temperature	400°C
Refractive index	2.62
Residual Stress	± 5MPa
Deposition rate	180nm/min
Uniformity (refractive index & thickness)	<1%

4.2. Pecvd A-Sic As Masking Layer

The intrinsic chemical inertness of SiC makes PECVD α-SiC an interesting candidate as masking layer for harsh wet and dry etching.

4.2.1. Masking Layer For Orientation Dependent Etching Of Si In Alkaline Solutions

A low etching rate of 78nm/h of low stress PECVDα-SiC in a 30%KOH solution was reported [19, 24,37], while etching rates lower than 2nm/h in both 33%KOH at 85°C as well as in 25% TMAH are reported by Sarro in [16]. As expected, the etching rate depends on both the composition of the deposited α-SiC layer and its density. However, in general, the reported results are in the same range with the etch rate values of PECVD Si_3N_4 (13 nm/h) in KOH 20% at 85°C [36] but much better than the etch rate of thermal SiO_2 (462 nm/h)in the same etchant [44]. As the reported etch rates of PECVD α-SiC have such relatively low values, we can conclude that PECVD α-SiC can be successfully used as a mask in silicon bulk micromachining processes. Furthermore, we have practically tested this hypothesis and demonstrated the feasibility of using PECVD α-SiC as masking layer, as is detailed in the next section.

4.2.2. Masking Layer For Etching In Hf Based Solutions

Deep wet etching of glass is an important technology for microfluidic applications [45]. The main etchant for glass materials is highly concentrated HF [46] (sometimes with a small amount of HCl [47] or H_3PO_4 added [48]). PECVD α-SiC is almost an inert material in these solutions, exhibiting etching rates lower than 10Å/h [37]. Such an extremely low etching rate in highly concentrated HF solutions combined with the reduced value of the stress are the main request of a good masking layer for deep wet etching of glasses [46, 49-51]. In our tests, a compound masking multilayer 'sandwich' made to low stress α-Si/low stress α-SiC/photoresist seemed to be the best solution in terms of depth of the etch achieved without any damage of the mask (through-etching of 1 mm thick Pyrex glass wafers, equivalent of a 2.5 hour exposure to 49% HF). The experimental results of processing Pyrex glass wafers (Corning 7740) in such a manner are presented in Fig. 6. It can be noticed that the mask is fully intact after the etching process while the shape of the etched hole describes a perfect isotropic process. The α-Si was used in this case as an adhesion layer. If only PECVD α-SiC is used as masking layer, a hugely isotropical etching process can still be observed [50].

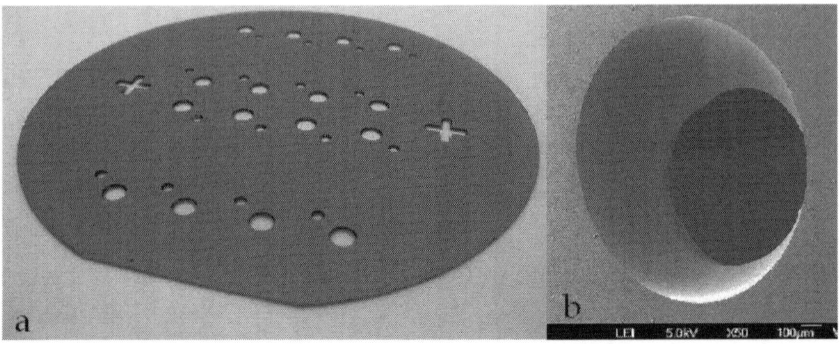

Figure 6. (a) Optical image of a glass wafer coated with PECVD α-Si/PECVD α-SiC/Photoresist after etching through in HF 49%, and (b) a SEM picture of one of the resulted etch-through holes in the glass wafer.

4.2.3. Masking Layer For Dry Release Structure In Xef_2

Dry release in XeF_2 is an emerging technology in surface and bulk micromachining of MEMS free-standing structures. The PECVD α-SiC films present a low etching rate in XeF_2 gas (around 7 Å/min) which makes the α-SiC very suitable as a structural layer for any dry-release processes using α-Si or polysilicon as a sacrificial layer. Fig. 7 presents SEM images with PECVD α-SiC cantilevers fabricated using dry released in XeF_2 [24].

4.2.4. Protective Layer For Harsh Environment

PECVD α-SiC films are also very suitable for structures intended to operate in harsh environments, due to α-SiC 's large hardness (2.48kg/m^2) [52], high fracture strength, high modulus [53], excellent wear resistance [54] and chemical inertness in acid or based solutions, low oxidation rate and strong covalent Si-C bonds [52]. Early work proved the potential of PECVD α-SiC as a potential material for encapsulation of micromachined transducers due to its good resistance in a large range of media such as piranha solution, HF and KOH [55].

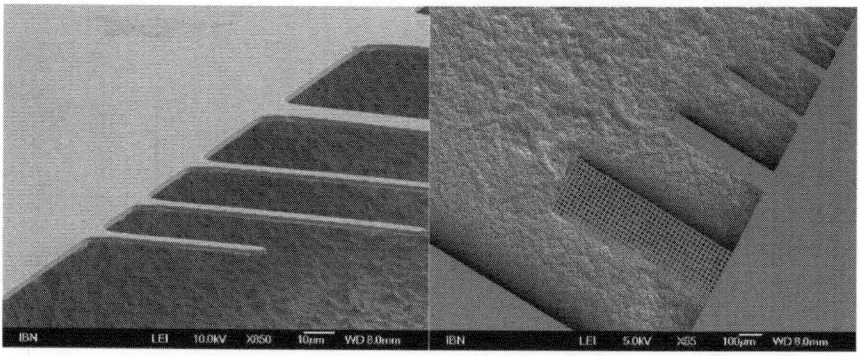

Figure 7. PECVD α-SiC cantilevers fabricated using dry-released process in XeF_2.

The good mechanical strength and anti-stiction surface properties of PECVD α-SiC [56] as well as its inertness in corrosive environment recommends this material for diverse applications. An illustrating example is that of a 1μm-thick PECVD α-SiC combined together with Teflon like fluoro-polymer coatings to reduce the demolding energy by a factor of about 10, compared to a bare silicon mold [57]. In another similar application PECVD α-SiC was used for its hardness and for its good step-coverage (improving the wall roughness generated during the deep RIE process) while the Teflon layer acted as an anti-stiction layer [58]. Decreasing the demolding energy, in this application, is equivalent with the increasing of the life-time of the Si mold (used usually for rapid prototyping on hot embossing tools).

In another application [30], a 1μm-thick PECVD α-SiC layer was used as an anti-erosion coating layer of a piezoresistive pressure sensor. In order to reduce the residual stress in the α-SiC protection layer (initially evaluated at −450MPa), annealing was performed at 450°C for 1h. The resulting stress value was of only +60MPa. The α-SiC protective layer also showed a good coverage of the Al metallization layer. The erosion testing was performed in 45%KOH solution at 80°C. The final PECVD α-SiC-coated pressure sensor showed a small decrease in sensitivity, but exhibited a high erosion resistance and less temperature dependence. A similar application was also reported [59], in which low stress PECVD α-SiC layer was used for a capacitive pressure sensor fabricated using surface micromachining. Aluminum was used as material for electrodes while polyimide was selected as a sacrificial layer.

4.3. Fabrication Of Free Standing Structuresof Pecvd A-Sic Using To Surface Micromachining

The opportunity of using thick α-Si (amorphous silicon) layers [22] (up to 20μm thick, from our own experience) in surface micromachining gives rise to new processing opportunities, especially for microfluidic applications. The α-Si sacrificial layer can be easily removed by wet etching in an alkaline solution (TMAH or KOH) or by dry release in XeF_2. The PECVD α-SiC presents a high chemical inertness to all the etchants of the above mentioned processes, and, therefore, is a very attractive candidate as a structural layer for surface micromachining processes in which amorphous silicon is used as a sacrificial layer. An example of such a process used a 3μm-thick free standing structure fabricated from PECVD α-SiC on a glass substrate using 9 μm-thick PECVD α-Si as sacrificial layer [22]. The structure was released using wet etching in 30% KOH at 80°C. The structure (presented in Fig. 8) shows also that the PECVD deposition process ensured a very good step coverage, the thickness of the vertical wall being identical with that of the layer on the horizontal surfaces [22]. In another example of using PECVD α-SiC as structural material for surface micromachining, 1μm-thick self-sustaining microbridges and microtunnels were fabricated from PECVD α-SiC films (deposited at 320°C using CH_4 and SiH_4 as reactive gases), using a SiO_xN_yfilm as a sacrificial layer [60]. In a similar manner, self-sustained grids have also been fabricated [61].

PECVD α-SiC was used in shunt capacitor RF MEMS microbridge-based switches [23]. In this application 300 nm and 500 nm thick PECVD α-SiC films were used as structural layers, while 2μm-thick PECVD SiO_2 was used as a sacrificial layer. The structures were released in BOE, followed by CO_2 supercritical drying in order to eliminate anystiction possibility. Polyimide was the choice for the sacrificial layer in an application which also used PECVD α-SiC as structural layer [21]. The main advantage of using polyimide is the opportunity of using a dry release process (in O_2 plasma), thus more easily overcoming stiction problems often encountered in wet sacrificial etching technology than by using XeF_2 or

CO_2 supercritical drying, solutions which require much more complex and expensive achiness.

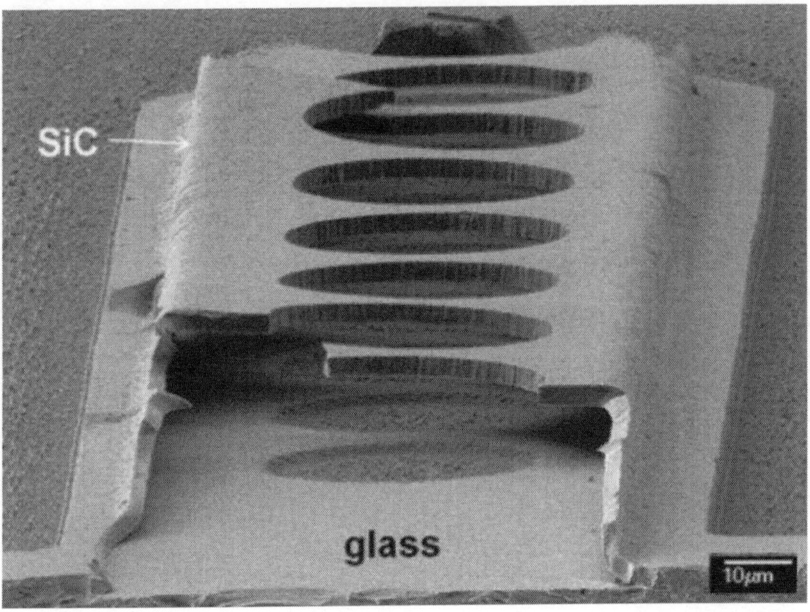

Figure 8. SEM image of asuspended structure fabricated by surface micromachining using thick low-stress α-Si as sacrificial layer and PECVD α-SiC as a structural layer [22].

4.4. Biological Applications Of Pecvd A-Sic

MEMS materials are suitable for applications in cell culture and tissue engineering [62]. In this direction, a special attention is given to the layers achieved by PECVD deposition especially α-Si_3N_4[63], [64-66] and α-SiC [37]. The main motivation in using these materials and PECVD as deposition method are related to:

1) Both α-Si_3N_4 and α-SiC can be used for the fabrication of 2-3μm thick membranes using classical micromachining processes.

2) PECVD deposition allows a good control of the residual stress in the layer, a critical aspect in achieving free standing structures.

3) Both α-Si_3N_4 and α-SiC are optically transparent, so the classical inverted microscopes use in biology can be easily used to monitor bio-samples in structures realized with these materials, for example, in assays emplying red/green fluorescence.

4) Using microfabrication, these membranes are attached to a supporting Si ring which makes the structure easy to be handle (comparing with polymeric membranes where no such rim support is present) [66].

5) Porous PECVD α-SiC and α-Si_3N_4 membranes can be fabricated with a good control of the pore size and with a uniform distribution of the pores can be achieved.

6) The commercially available and presently used polymeric membranes are hydrophobic materials, and metabolites (albumin, urea) are absorbed in this membrane with effect on cell viability.

7) The thickness of polymeric porous membrane is usually ranging from tens to hundreds of microns. Consequently, the mass transfer of (bio) chemical substances across/through the membrane is strongly affected by the thickness of the membrane and its nature (hydrophilic/hydrophobic).

8) The presence of functional groups, such as amine (-NH_2), and methyl (-CH_3), on the surface of α-Si_3N_4 and α-SiC membranes can significantly enhance the cell adhesion on these materials.

9) The α-SiC and α-Si_3N_4 membranes can be easily cleaned: complete removal of all organic contaminants can be achieved in piranha solution, followed by subsequent rinsing, drying and if necessary even sterilization can also be performed, in order to ensure complete reusability.

As an illustrative example, chips with 2.5μm-thick PECVD α-SiC membranes were fabricated by bulk micromachining in order to test the biocompatibility of the α-SiC [37]. Fibroblast NIH3T3 cells were used in this study as the model cell line. It was observed that the presence of CH_3 groups on the surface of the membranes improved the cell adhesion. Moreover, dipping for 1 minute in 40% NH_4F, which was performed mainly to reduce the density of native silicon oxide groups on the α-SiC surface, also improved the adhesion of the cells on the membranes' surfaces. Fig. 9 shows cell culture images taken 24 hours and 48 hours after starting the culturing.

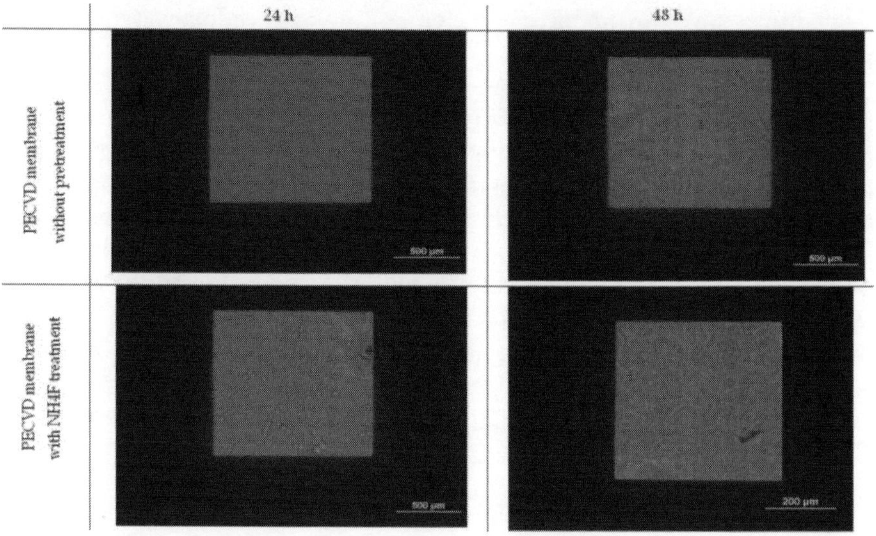

Figure 9. Optical images with the adhesion of fibroblast NIH3T3 cells on PECVD α-SiC membrane treated in NH_4F after 24h and 48h cell culture.

5. CONCLUDING REMARKS

We can conclude that PECVD α-SiC is a very attractive and promising material. Its PECVD deposition enable low temperature processing and thus ensures compatibility with CMOS processing, which is essential when (bio) MEMS/NEMS structures need to be combined together with signal processing and data conditioning circuits, a *sine qua non* condition of realizing "smart sensors". Additionally, PECVD deposition enables the user to modify various deposition parameters (temperature, pressure, precursors gas flow ratios) which can allow the user to optimize –at least in principle- the desired film properties: chemical composition internal average stress and refractive index. However, in practice such multi-dimensional optimization may prove difficult to achieve, and the presence of some extra degrees of freedom is welcome. An example to the point is -in our case- the presence of dual LF and HF modes for the STS Multiplex Pro-CVD PECVD reactor, which provided valuable flexibility in usage and in varying independently various characteristics of the film (e.g. minimizing the stress) without compromising the others (e.g. the refractive index). In some cases, when such a machine is not available, some more maneuvering room is given by adding an anneal process, which always reduces (sometimes significantly) the film's stress.

Application-wise, the long list of extremely desirable properties exhibited by α-SiC (large hardness, high fracture strength, high modulus, excellent wear resistance, superb chemical inertness and excellent stability even at high temperatures and/or under corrosive conditions) ensure it can be used in a very large area of applications, particularly in harsh environments, where Si devices cannot be used. Moreover, the same properties recommend PECVD α-SiC as the material of choice for a large range of MEMS postprocessing fabrication techniques. Thus, it can be used as mold material (or mold-coating protective layer), masking material or active structural material in surface micromachining.

Finally, we should also highlight that α-SiC can has also been employed in bio-related applications. Our own tests have shown that PECVD α-SiC membranes fabricated by bulk micromachining can be used successfully in cell cultures. Here the intrinsic properties of α-SiC ensure very good chemical stability, possibility to easily functionalize the surface for, e.g., increased cell adhesion, and improved mechanical resistance and easy handling.

REFERENCES

1. W. M. Zhou, F. Fang, Z. Y. Hou, L. J. Yan, Y. F. Zhang, 2006 IEEE Electron Dev. Lett. 27 6 463 465

2. C. Patil, F. Xiao-An, C. Anupongongarch, M. Mehregany, S. Garverick, 2007 Proc. IEEE Compound Semicond. Int. Circuit Symp. 1 4

3. N. Sarazin, E. Morvan, M. A. di Forte Poisson, M. Oualli, C. Gaquiere, O. Jardel, O. Drisse, M. Tordjman, M. Magis, S. L. Delage, 2010 IEEE Electron Dev. Lett. 31 1 11 13

4. D. Peftitsis, G. Tolstoy, A. Antonopoulos, J. Rabkowski, L. Jang-Kwon, M. Bakowski, H. Nee, 2012 IEEE Transactions on Power Electr. 27 1 28 36

5. S. Woongje, E. Van Brunt, B. J. Baliga, A. Q. Huang, 2011 IEEE Electron Dev.Lett. 32 7 880 882

6. V. Nikiforov, A. L. Tomás García, I. M. Petrushina, E. Christensen, N. J. Bjerrum, 2011 Int. J. Hydrogen Energy 36 10 5797 5805

7. N. B. Singh, B. Wagner, A. Berghmans, D. J. Knuteson, S. Mc Laughlin, D. Kahler, D. Thomson, M. King, 2010 Crystal Growth & Design 10 8 3508 3514

8. D. R. Myers, K. B. Cheng, B. Jamshidi, R. G. Azevedo, D. G. Senesky, L. Chen, M. Mehregany, M. B. J. Wijesundara, A. P. Pisano, 2009 J. Micro-Nanolithogr. MEMS MOEMS 8 2

9. M. T. Soo, K. Y. Cheong, A. F. M. Noor, 2010 Sens. Actuator B-Chem. 151 1 39 55

10. E. Borchi, R. Macii, M. Bruzzi, M. Scaringella, 2011 Nuclear Instruments and Methods in Physics Research Section A: Accelerators, Spectrometers, Detectors and Associated Equipment 658 1 121 124

11. G. R. Azevedo, G. D. Jones, V. A. Jog, B. Jamshidi, R. D. Myers, L. Chen, Fu X.-a., M. Mehregany, B. J. M. Wijesundara, P. A. Pisano, 2008 Sens. Actuator A-Phys. 145-146 0 2 8

12. L. Jiang, R. Cheung, J. Hedley, M. Hassan, A. J. Harris, J. S. Burdess, M. Mehregany, C. A. Zorman, 2006 Sens. Actuator A-Phys. 1128 2 376 386

13. J. S. Hyun, J. H. Park, J. S. Moon, S. H. Kim, Y. J. Choi, N. E. Lee, J. H. Boo, 2005 Progress in Solid State Chemistry 33 2-4 309 315

14. A. Zorman, M. Mehregany, 2002 Proc. of IEEE Sensors 2 1109 1114

15. F. Liu, C. Carraro, A. P. Pisano, R. Maboudian, 2010 J. Micromech. Microeng. 20 3

16. P. M. Sarro, C. R. Deboer, E. Korkmaz, J. M. W. Laros, 1998 Sens. Actuator A-Phys. 67 1-3 175 180

17. M. A. El Khakani, M. Chaker, A. Jean, S. Boily, J. C. Kieffer, M. E. O'Hern, M. F. Ravet, F. Rousseaux, 1994 J. Mat. Research 9 1 96 103

18. S. Roper, R. T. Howe, R. Maboudian, 2006 J. Micromech. Microeng. 16 12 2736 2739

19. M. Iliescu, B. Avram, A. Chen, V. Popescu, D. P. Dumitrescu, A. Poenar, D. Sterian, S. Vrtacnik, P. Amon, Sterian, 2011 J. Optoel. Adv. Mat. 13 4 387 394

20. P. M. Sarro, 2000 Sens. Actuator A-Phys. 82 1-3 210 218

21. L. Bagolini, T. L. M. Pakula, H. T. M. Scholtes, P. J. Pham, P. M. French, Sarro, 2002 J. Micromech. Microeng. 12 4 385 389

22. B. Iliescu, T. Chen, 2008 J. Micromech. Microeng. 18 1 15024

23. R. J. Parro, M. C. Scardelletti, N. C. Varaljay, S. Zimmerman, C. A. Zorman, 2008 Solid-State Electronics 52 10 1647 1651

24. C. Iliescu, B. T. Chen, J. S. Wei, A. J. Pang, 2008 Thin Solid Films 516 16 5189 5193

25. M. Avram, A. Avram, A. Bragaru, B. Chen, D. P. Poenar, C. Iliescu, 2010 Proc. of the 33rdIEEE Int. Semicond. Conf. 239 242

26. H. T. M. Pham, C. R. De Boer, K. Kwakernaak, W. G. Sloof, P. M. Sarro, 2001 Proc. of SPIE 272 279

27. R. Oliveira, M. N. P. Carreno, 2006 J. Non-Crystalline Solids 352 9-20 1392 1397

28. G. Pandraud, P. J. French, P. M. Sarro, 2008 Sens. Actuator A-Phys. 142 1 61 66

29. G. Pandraud, P. J. French, P. M. Sarro, 2005 Microw. Opt. Technol. Lett. 47 3 219 220

30. H. Zhang, H. Guo, Y. Wang, G. Zhang, Z. Li, 2007 J. Micromech. Microeng. 17 3 426 431

31. G. A. Slack, S. F. Bartram, 1975 J. Appl. Phys. 46 1 89 98

32. R. G. Azevedo, Z. Jingchun, D. G. Jones, D. R. Myers, A. V. Jog, B. Jamshidi, M. B. J. Wijesundara, R. Maboudian, A. P. Pisano, 2007 Proc. of IEEE 20th Int. Conf. on Micro Electro Mechanical Systems, MEMS. 643 646

33. V. I. Ivashchenko, S. N. Dub, O. K. Porada, L. A. Ivashchenko, P. L. Skrynskyy, A. I. Stegniy, 2006 Surf. Coat. Technol. 200 22-23 6533 6537

34. N. Behnel, T. Fuchs, H. Seidel, 2009 Proc. of Int. Conf. on Solid-State Sens., Act. and Microsyst., TRANSDUCERS 2009 740 742

35. C. K. Chung, M. Q. Tsai, P. H. Tsai, C. Lee, 2005 J. Micromech. Microeng. 15 1 136 142

36. C. Iliescu, F. E. H. Tay, J. S. Wei, 2006 J. Micromech. Microeng. 16 4 869 874

37. C. Iliescu, B. Chen, D. P. Poenar, Y. Y. Lee, 2008 Sens. Actuator B-Chem. 129 1 404 411

38. S. Tanaka, K. Rajanna, T. Abe, M. Esashi, 2001 J. Vac. Sci. Technol. B 19 6 2173 2176

39. P. Chabert, 2001 J. Vac. Sci. Technol. B 19 4 1339 1345

40. U. Schmid, M. Eickhoff, C. Richter, G. Krotz, D. Schmitt-Landsiedel, 2001 Sens. Actuator A-Phys. 94 1-2 87 94

41. M. B. J. Gao, C. Wijesundara, R. T. Carraro, Howe, R. Maboudian, 2004 IEEE Sens. J. 4 4 441 448

42. D. G. Senesky, A. P. Pisano, 2010 Proc. of 23rd IEEE Int. Conf. on Micro Electro Mechanical Systems (MEMS) 352 355

43. H. T. M. Pandraud, P. J. Pham, P. M. French, Sarro, 2007 Optics & Laser Technology 39 3 532 536

44. K. R. Williams, K. Gupta, M. Wasilik, 2003 J. Microelectromech. Syst. 12 6 761 778

45. C. Iliescu, H. Taylor, M. Avram, J. Miao, S. Franssila, 2012 Biomicrofluidics 6 1 016505 016516

46. C. Iliescu, B. Chen, J. Miao, 2008 Sens. Actuator A- Phys. 143 1 154 161

47. C. Iliescu, J. Jing, F. E. H. Tay, J. Miao, T. Sun, 2005 Surf. Coat. Tech. 198 1-3 314 318

48. F. Berthold, H. Laugere, C. R. Schellevis, M. De Boer, R. M. Laros, P. M. Guijt, M. J. Sarro, 2002 Vellekoop Electrophoresis 23 20 3511 3519

49. F. E. H. Iliescu, J. M. Tay, Miao, 2007 Sens. Actuator A-Phys. 133 2 395 400

50. X. Zhang, H. Guo, Z. Chen, G. B. Zhang, Z. H. Li, 2007 J. Micromech. Microeng. 17 4 775 780

51. P. Poenar, C. Iliescu, M. Carp, A. J. Pang, K. J. Leck, 2007 Sens. Actuator A-Phys. 139 1-2 162 171

52. A. Zorman, A. C. Barnes, 2012 in Silicon Carbide Biotechnology Elsevier Oxford 351 376

53. W. N. Sharpe, O. Jadaan, G. M. Beheim, G. D. Quinn, N. N. Nemeth, 2005 J. Microelectromech. Syst. 14 5 903 913.

54. W. R. Ashurst, M. B. J. Wijesundara, C. Carraro, R. Maboudian, 2004 Tribology Letters 17 2 195 198

55. F. Flannery, N. J. Mourlas, C. W. Storment, S. Tsai, S. H. Tan, J. Heck, D. Monk, T. Kim, B. Gogoi, G. T. A. Kovacs, 1998 Sens. Actuator A-Phys. 70 1-2 48 55

56. J. B. Summers, M. Scardelletti, R. Parro, C. A. Zorman, 2007 in MEMS/MOEMS Components and Their Applications IV, edited by S. A. Tadigadapa, R. Ghodssi and A. K. Henning Proc. Spie-Int Soc Optical Engineering 6464 H4640 H4640

57. D. Taylor, C. Boning, Iliescu, 2011 J. Micromech. Microeng. 21 6

58. X. Gao, L. P. Yeo, M. B. Chan-Park, J. M. Miao, Y. H. Yan, J. B. Sun, Y. C. Lam, C. Y. Yue, 2006 J. Microelectromech. Syst. 15 1 84 93

59. L. S. Pakula, H. Yang, H. T. M. Pham, P. J. French, P. M. Sarro, 2004 J. Micromech. Microeng. 14 11 1478 1483

60. M. N. P. Carreño, A. T. Lopes, 2004 J. Non-Crystalline Solids 338-340 0 490 495

61. M. N. P. Carreno, M. I. Alayo, I. Pereyra, A. T. Lopes, 2002 Sens. Actuator A-Phys. 100 2-3 295 300

62. M. Ni, W. H. Tong, D. Choudhury, N. A. A. Rahim, C. Iliescu, H. Yu, 2009 Int. J. Mol. Sci. 10 12 5411 5441

63. S. Zhang, W. Tong, B. Zheng, T. A. K. Susanto, L. Xia, C. Zhang, A. Ananthanarayanan, X. Tuo, R. B. Sakban, R. Jia, C. Iliescu, K. , H. Chai, M. Mc Millian, S. Shen, H. Leo, H. Yu, 2011 Biomaterials 32 4 1229 1241

64. D. R. Ciarlo, 2002 Biomed. Microdev. 4 1 63 68

65. S. Wei, J. Kwong, P. Gaughwin, M. Avram, C. Iliescu, 2009 Int. J. Comp. Mat. Science Surface Eng. 2 3-4 268 281

66. C. Iliescu, J. Wei, B. Chen, P. L. Ong, 2008 Rom. J. Inform. Sci. Technol. 11 2 167 176

INDEX

Q

R

S

W